LOCUST OUTBREAKS

Management and the World Economy

LOCUST OUTBREAKS

Management and the World Economy

Edited by

Umair Riaz, PhD

Khalid Rehman Hakeem, PhD

AAP | APPLE ACADEMIC PRESS

First edition published 2024

Apple Academic Press Inc.
1265 Goldenrod Circle, NE,
Palm Bay, FL 32905 USA

760 Laurentian Drive, Unit 19,
Burlington, ON L7N 0A4, CANADA

CRC Press
6000 Broken Sound Parkway NW,
Suite 300, Boca Raton, FL 33487-2742 USA

4 Park Square, Milton Park,
Abingdon, Oxon, OX14 4RN UK

Apple Academic Press exclusively co-publishes with CRC Press, an imprint of Taylor & Francis Group, LLC

Library and Archives Canada Cataloguing in Publication

Title: Locust outbreaks : management and the world economy / edited by Umair Riaz, PhD, Khalid Rehman Hakeem, PhD.
Names: Riaz, Umair, editor. | Hakeem, Khalid Rehman, editor.
Description: First edition. | Includes bibliographical references and index.
Identifiers: Canadiana (print) 20230171478 | Canadiana (ebook) 20230171524 | ISBN 9781774912331 (hardcover) | ISBN 9781774912348 (softcover) | ISBN 9781003336716 (ebook)
Subjects: LCSH: Locusts—Control. | LCSH: Agricultural pests—Control. | LCSH: Locusts—Economic aspects.
Classification: LCC SB945.L7 L63 2023 | DDC 632/.726—dc23

Library of Congress Cataloging-in-Publication Data

Names: Riaz, Umair, editor. | Hakeem, Khalid Rehman, editor.

Title: Locust outbreaks : management and the world economy / edited by Umair Riaz, PhD, Khalid Rehman Hakeem, PhD.

Description: First edition. | Paln Bay, FL : Apple Academic Press, 2023. | Includes bibliographical references and index. | Summary: "This book is the comprehensive resource on the devastating effect of locust outbreaks on crop production along with valuable coverage of traditional as well as new and emerging control and mitigation strategies. Locust Outbreaks: Management and the World Economy begins with an introduction to and history of locust attacks and addresses the deleterious effects of locusts on crops. The chapters delve into both traditional and state-of-the art control and management technologies to combat locust outbreaks, including such advanced technologies as geographical information systems (GIS) and global positioning systems (GPS) and methods such as employing entomopathogenic fungi as a pest control. The book also evaluates the damage of locusts due to climate change and its environmental ramifications. It considers the scientific aspect of the role of pheromones on the reproduction of locusts and discusses the culture of using locusts as a food source in some countries. Topically, the volume also considers the influence of the COVID-19 pandemic on locust outbreaks on the world's economy. This important and unique volume is a highly valuable resource for those working on meeting the challenges of locust invasions and their terrible consequences on crops, societies, and economies"-- Provided by publisher.

Identifiers: LCCN 2023006325 (print) | LCCN 2023006326 (ebook) | ISBN 9781774912331 (hardback) | ISBN 9781774912348 (paperback) | ISBN 9781003336716 (ebook)

Subjects: LCSH: Locusts--Control.

Classification: LCC SB945.L7 L63 2023 (print) | LCC SB945.L7 (ebook) | DDC 632.726--dc23/eng/20230322

LC record available at https://lccn.loc.gov/2023006325

LC ebook record available at https://lccn.loc.gov/2023006326

ISBN: 978-1-77491-233-1 (hbk)
ISBN: 978-1-77491-234-8 (pbk)
ISBN: 978-1-00333-671-6 (ebk)

Dedication

This book is dedicated to my beloved Mother ***"Hajrah Begum"***, who left this world so early. To me, she was the whole world. She was the most loving, humble, compassionate, understanding, family-orientated woman. I am going to miss you so much, ***Mom***. I pray to Allah (SWT) to bless your soul and grant you, Jannatul-Firdous (the highest level of heaven).

'O Allah, forgive and have mercy upon her, excuse her and pardon her, and make honourable her reception. Expand her entry, cleanse her with water, snow, and ice, and purify her of sin as a white robe is purified of filth. Exchange her home for a better home. Admit her into the Garden; protect her from the punishment of the grave and the torment of the Fire'.

—Prof. (Dr.) Khalid Rehman Hakeem

About the Editors

Umair Riaz, PhD

Umair Riaz, PhD, is working as an Assistant Professor in the Department of Soil & Environmental Sciences at MNS University of Agriculture Multan, Pakistan. Dr. Riaz also served as a Scientific Officer (Lab) for six years in the Soil and Water Testing Laboratory for Research, Bahawalpur, Pakistan, specializing in waste management, metal toxicology, phytochemistry, and climate change. Dr. Riaz's research interests are related to the development of new fertilizer technology based on nanochemistry with an emphasis on plant nutrition. He has supervised graduate and postgraduate students of environmental sciences for the past several years. He is the author of more than 50 research papers and book chapters, and he has presented and participated in numerous state, national, and international conferences, seminars, workshops, and symposia. Dr. Riaz has worked as a research associate in Higher Education Commission (HEC)-funded projects regarding field studies. He has received many awards, appreciations, and recognitions for his services to the science of soil, water, pesticide, and fertilizer testing analysis. He has also served as an editorial board member and reviewer of international journals.

Khalid Rehman Hakeem, PhD

Khalid Rehman Hakeem, PhD, is Professor at King Abdulaziz University, Jeddah, Saudi Arabia. After completing his doctorate (botany; specialization in plant eco-physiology and molecular biology) from Jamia Hamdard, New Delhi, India, he worked as a lecturer at the University of Kashmir, Srinagar, India, at Universiti Putra Malaysia, Selangor, Malaysia, he was a Postdoctorate Fellow and Fellow Researcher (Associate Professor) for several years. Dr. Hakeem has more than 14 years of teaching and research experience in plant eco-physiology, biotechnology and molecular biology, medicinal plant research, plant-microbe-soil interactions, as well as in environmental studies. He is the recipient of several fellowships at both national and international levels. He has served as a visiting scientist at Jinan University, Guangzhou, China.

Currently, he is involved with a number of international research projects with different government organizations. To date, Dr. Hakeem has authored and edited more than 85 books with international publishers. He also has to his credit more than 180 research publications in peer-reviewed international journals and 55 book chapters in edited volumes with international publishers. At present, Dr. Hakeem serves as an editorial board member and reviewer for several high-impact international scientific journals. He is included in the advisory board of Cambridge Scholars Publishing, UK. He is fellow of Royal Society of Biology (UK), fellow of the Plantae group of the American Society of Plant Biologists, member of the World Academy of Sciences, member of the International Society for Development and Sustainability, Japan, and member of the Asian Federation of Biotechnology, Korea.

Contents

Contributors

Rayees Ahmad
Entomology Research Lab, Department of Zoology, University of Kashmir, Jammu & Kashmir, India

Tariq Ahmad
Department of Zoology, University of Kashmir, Srinagar, Jammu & Kashmir, India

Imran Akhter
Regional Agricultural Research Institute, Bahawalpur, Pakistan

M. Ali
Department of Pharmacognosy, College of Pharmacy, Jazan University, Jazan, Kingdom of Saudi Arabia

Ayesha Amir
Sustainable Development Study Center, Government College University, Lahore, Pakistan

Wajiha Anum
Regional Agricultural Research Institute, Bahawalpur, Pakistan

Aqsa
Department of Soil Science, University of Agriculture Faisalabad, Sub Campus Burewala-Vehari, Pakistan

Mustansar Aslam
Department of Agricultural Sciences, Institute of Environmental & Agricultural Science, University of Okara, Okara, Pakistan

Tabinda Athar
Institute of Soil and Environmental Sciences, University of Agriculture Faisalabad, Faisalabad, Pakistan

Uzma Azeem
Sanmati Government College of Science Education and Research, Jagraon, Ludhiana, Punjab, India

Humera Aziz
Department of Environmental Sciences and Engineering, Government College University, Faisalabad, Pakistan

Muhammad Farhan
Sustainable Development Study Center, Government College University, Lahore, Pakistan

Sana Farooq
Department of Environmental Sciences and Engineering, Government College University, Faisalabad, Pakistan

Hina Fatima
School of Applied Biosciences, Kyungpook National University, Daegu, South Korea

Muhammad Fiaz
Institute of Plant Protection, MNS University of Agriculture, Multan, Pakistan

M. Usman Ghazanfar
College of Agriculture, University of Sargodha, Pakistan

Khalid Rehman Hakeem
Department of Biological Sciences, King Abdul Aziz University, Jeddah, Kingdom of Saudi Arabia

Ayesha Hassan
Department of Environmental Sciences, Faculty of Natural Sciences, University of Okara, Okara, Pakistan

Shafia Hassan
Department of Zoology, University of Kashmir, Srinagar, Jammu & Kashmir, India

David Hunter
Orthopterists' Society, McKellar, ACT 2617, Australia

Barkat Hussain
Division of Entomology, Department of Horticulture, Srinagar, Jammu & Kashmir, India

Muhammad Ishtiaq
Institute of Plant Protection, MNS University of Agriculture, Multan, Pakistan

Naeem Iqbal
Institute of Plant Protection, MNS University of Agriculture, Multan, Pakistan

Amina Kanwal
Department of Botany, Government College Women University, Sialkot, Pakistan

Nafisa Kanwal
Department of Soil Science, University of Agriculture Faisalabad, Sub Campus Burewala-Vehari, Pakistan

Arslan Khan
Institute of Plant Protection, MNS University of Agriculture, Multan, Pakistan

Madiha Mobeen Khan
Regional Agricultural Research Institute, Bahawalpur, Pakistan

Naeem Arshad Maan
Regional Agricultural Research Institute, Bahawalpur, Pakistan

Farah Majid
Department of Entomology, University of Agriculture, Faisalabad, Faisalabad, Pakistan

Muhammad Aqeed Mehdi
Institute of Plant Protection, MNS University of Agriculture, Multan, Pakistan

Khair-Ul-Nisa
Department of Environmental Science, University of Kashmir, Srinagar, India

Mirza Abdul Qayyum
Institute of Plant Protection, MNS. University of Agriculture, Multan, Pakistan

Hasan Riaz
Institute of Plant Protection, MNS University of Agriculture, Multan, Pakistan

Umair Riaz
Department of Soil & Environmental Sciences, MNS University of Agriculture Multan, Pakistan

Anjum Sabba
Department of Biochemistry, University of Kashmir, Srinagar, India

Shafqat Saeed
Institute of Plant Protection, MNS University of Agriculture, Multan, Pakistan

Saliqa Salwat
Division of Entomology, SKUAST-K, Srinagar, Jammu & Kashmir, India

Laila Shahzad
Sustainable Development Study Center, Government College University, Lahore, Pakistan

Muhammad Umair Sial
Institute of Plant Protection, MNS University of Agriculture, Multan, Pakistan

Aboubakar Siddique
Institute of Plant Protection, MNS University of Agriculture, Multan, Pakistan

Zahra Sultan
Department of Botany, University of Kashmir, Srinagar, India

Najeebul Tarfeen
Centre of Research for Development (CORD), University of Kashmir, Srinagar, Jammu & Kashmir, India

Unsar Naeem-Ullah
Institute of Plant Protection, MNS University of Agriculture, Multan, Pakistan

Muhammad Wajid
Institute of Plant Protection, MNS. University of Agriculture, Multan, Pakistan

Waqas Wakil
Department of Entomology, University of Agriculture, Faisalabad, Pakistan
Department of Continuing Education, University of Agriculture, Faisalabad, Pakistan

Saba Wani
Department of Biochemistry, University of Kashmir, Srinagar, India

Aisha A. Waris
Institute of Soil and Environmental Sciences, University of Agriculture Faisalabad, Faisalabad, Pakistan

Asma Yasin
Sustainable Development Study Center, Government College University, Lahore, Pakistan

Muhammad Yasin
Department of Entomology, The Islamia University, Bahawalpur, Pakistan

Abbreviations

ADB	Asian Development Bank
CIDA	Canadian International Development Agency
CSPM	climate-smart pest management
DL	desert locusts
DFID	Department for International Development, UK
DGIS	Dutch Directorate General of International Cooperation
DLIS	Desert Locust Information Service
FAO	Food and Agriculture Organization
GAM	Global Agricultural Monitoring
GPS	Global Positioning Systems
GIS	geographical information system
HtA	hirsutellin A
ICTs	Information and Communication Technologies
IIBC	International Institute of Biological Control
IPM	integrated pest management
JA	jasmonic acid
JH	juvenile hormone
LUCC	land use and cover change
MERS	Middle East Respiratory Syndrome
MODIS	Moderate Resolution Image Spectroradiometer
NGOs	nongovernmental organizations
NSKE	neem seed kernel extract
NVDI	normalized difference vegetation index
OLI	Operational Land Imager
PAN	phenylacetonitrile
PCS	preventive control strategy
PDI	proportional differential index
RAMSES	Reconnaissance and Monitoring System of the Environment of Schistocerca
RS	remote sensing
SARS	Severe Acute Respiratory Syndrome
SDC	Swiss Agency for Development and Cooperation
SWARMS	Schistocerca Warning and Management System

TPI	Transition Probability Index
ULV	ultra-low volume
VOCs	volatile organic compounds
WHO	World Health Organization

Preface

Agriculture, the backbone of economy, has been put under tremendous pressure by various global threats like the COVID-19 pandemic, locust swarms, and tsunami floods in various regions of the world. The developing countries have no capacity to bear these losses, while many precautionary measures, mitigation strategies, and control measures and practices are available in the developed countries. Locust outbreaks are directly linked with climate change, sea level rise, topography, latitude, temperature, and rainfall. Extraordinary weather and climate circumstances, especially extensive and severe rains since October 2019, have contributed to a serious and widespread locust outbreak. Arid and semiarid areas with minimum rainfalls undergo water shortage and crops flourish at risk. In addition the desert locust attacks result in social-economic and environmental disturbances over prolonged periods. In 1980, 1998, and 2010, locust outbreaks occurred in Africa, Australia, and New South Wales, respectively. These were the major outbreaks in recent history which destroyed more than a 390,000 km2 area. This count was about 50% of cultivated area, around a 2 billion US$ loss. In different countries/customs, locusts are used for eating in a variety of methods, like boiling, toasting, roasting, frying, sun-drying, or even raw. Their flavor is similar to walnuts and almonds.

This current book covers traditional as well as modern methods that are followed in different African and Asian countries. It also covers the new technologies to be used for the locust control and management such as geographical information systems (GIS) and global positioning systems (GPS). Robust policy interpolation is needed to control this havoc in coming times.

The book is an accurate and broad description of the involvement of different strategies for locust control and the sustainable strategies for healthy environs. Academicians, researchers, and students will find it a perfect resource on "locust outbreak and agriculture" as intrusion in agricultural practices for the reclamation of degraded soil environs and should sufficiently suffice the requirements of training, teaching, and research purposes.

Umair Riaz, Multan, Pakistan

Khalid Rehman Hakeem, Jeddah, Saudi Arabia

Introduction

Locust swarms threaten the world population with respect to food security. This issue has the potential of a disaster equal to a nuclear bomb. Large desert locust outbreaks are reported in literature over the years; these outbreaks are directly linked with warm and dry climates. Extraordinary weather and climate circumstances, especially extensive and severe rains since October 2019, have contributed to a serious and widespread locust infestation. As per FAO record, the year 2020–2021 outbreak was one of the deadliest in 70 years, which destroyed 70,000 ha land in Ethiopia, Somalia, and South Asia.

A typical swarm can have as many as 150 million locusts per square kilometer and can travel up to 150 km in a single day. Even a small locust swarm of 1 square kilometer will consume the same amount of food as 35,000 people in a single day. As a result, a swarm of desert locusts of 40–80 million locusts in less than 1 square kilometer will consume roughly 190 million kilograms of plants per day. The desert locust plagues are estimated to threaten the economic livelihood of one-tenth of the world's population. Devastating natural disasters caused by locust plagues wreak havoc on agriculture in many parts of the world, especially in Africa. Morocco, for example, lost more than $50 million in one season due to desert locust attacks, and Ethiopia lost 167,000 tons of grain in 1958, enough to feed 1 million people for a year.

Several techniques were employed to manage locusts' swarms. The basic idea evolving since the colonial times was to completely destroy their breeding grounds before they could again become a threat. A very popular technique was to use the Cyprus screens that were oil-tarred screens for killing locusts, but it was proved to be not so effective later on. Other popular methods were the use of the net system and the dhotar method. The first incorporates a capacious bag that was swung around the fields to trap young locusts. The latter involves using a blanket to trip locusts resting on bushes. The main limitation of such orthodox techniques was that it demands huge manpower. However, the majority of Indians have a mindset that these attacks are a “heaven-sent visitation”, and they think that their god would take care of it.

This book covers all the aspects of locusts, including historical background, life cycle, taxonomy, swarm history, and control measures (remote sensing, use of latest monitoring technologies, biological control, biorational control, insect growth regulators, genetic control, integrated pest management, etc.).

CHAPTER 1

Introduction, Historical Background, and Current Status of Desert Locusts

MUHAMMAD FARHAN[1] and AMINA KANWAL[2]

[1]Sustainable Development Study Center, Government College University, Lahore, Pakistan

[2]Department of Botany, Government College Women University, Sialkot, Pakistan

ABSTRACT

Locusts belong to the family Acrididae and are included in the subgroups of short-horned grasshoppers, they have swarming phase. Usually, they are solitary, but with the increase in number, they can change their habit and behavior and can become gregarious (Table 1.1). Locusts are usually innocuous, especially when their member/density is low and do not pose major damage to crops (Crook et al., 2020). However, under drought conditions and followed by rapid vegetation growth, their number begins to increase rapidly. This is due to the serotonin (secreted by brain) which imparts drastic set of changes in their body, their breeding increases their number abundantly, and they start becoming nomadic and gregarious. They form colonies of wingless nymphs which later develop into winged adults (Yang et al., 2020). They can fly to greater distances and have a tendency to consume most of the green vegetations whatever comes in their way or whereever the swarms settle. They are responsible for causing plagues, famines, and human migrations since pre-history. These are mentioned in

Locust Outbreaks: Management and the World Economy. Umair Riaz, Khalid Rehman Hakeem, (Eds.)

the Bible, the Mahabharata, and the Iliad. Locusts were carved on ancient Egyptians tombs. Recent developments in agricultural practices have enabled us to have better control over locusts. They are edible, and in many countries, they are expensive, rare, and highly desirable (Zhen et al., 2020).

1.1 LOCUST LIFE CYCLE

The life cycle of locusts can be grouped into three categories (Fontana et al., 2017):

a) Polyvoltine cycle (having multiple generations in 1 year).
b) Univoltine cycle (having one generation in 1 year).
c) Biennial and triennial cycle.

The life cycle of the desert locust consists of three distinct stages (Veran et al., 2015):

a) Embryonic development
b) Hopper or post-embryonic development
c) Adults

After fertilization, the female digs a chamber in soil with the help of abdomen and hind legs. She then lays eggs at the bottom of the chamber, the chamber is then covered with soil. The opening of the chamber is the covered with a waterproof cap, this cap protects the eggs from harsh climate and from other enemies. Usually, there is one generation per year. The larva (vermiform larva) is covered with semi-transparent, colorless, thin membrane, which is made up of cuticle (Branson, 2017).

TABLE 1.1 List of Locust Species.

Locust species	Area	Description
Aiolopus simulatrix	Sudan	Sudan plague locust
Anacridium aegyptium	Central Asia, northern Africa and Europe	Egyptian locust
Anacridium melanorhodon	Africa	Sahelian tree locust
Anacridium wernerellum	Sudan	Sudan tree locust
Austracris guttulosa	Australia	Spur-throated locust
Austroicetes cruciata	Australia	Small plague locust

TABLE 1.1 *(Continued)*

Locust species	Area	Description
Calliptamus italicus	Central Asia, central Europe and Morocco	Italian locust
Ceracris kiangsu	Southern China	Yellow-spined bamboo locust
Chortoicetes terminifera	Australia	Australian plague locust
Dissosteira longipennis	North America	High plains locust
Dociostaurus maroccanus	western Asia	Moroccan locust
Gastrimargus musicus	Australia	Yellow-winged locust
Gomphocerus sibiricus	Europe	Siberian locust
Locusta migratoria	Africa, Asia. Eastern Europe	Migratory locust
Locustana pardalina	Southern Africa	Brown locust
Melanoplus differentialis	Canada	Differential locust
Melanoplus sanguinipes	Caribbean	Migratory grasshopper
Melanoplus spretus	North America	Rocky Mountain locust
Nomadacris septemfasciata	Southern Africa	Red locust
Nomadacris succincta	India	Bombay locust
Oedaleus senegalensis	Canary Islands	Senegalese locust
Rhammatocerus schistocercoides	Brazil	Mato Grosso locust
Schistocerca cancellata	South America	South American locust
Schistocerca gregaria	Western India	Desert locust
Schistocerca piceifrons	Central America	Central American locust
Schistocerca interrita	America	Peru locust
Trimerotropis pallidipennis	North America	Pallid-winged locust

Source: Steedman (1988), Pener and Simpson (2009), Wikipedia (2020).

1.2 PHASE POLYPHENISM

Locusts show remarkable phase polyphenism, which is a phenotypic plasticity where locust can change their morphological, physiological, and behavioral traits due to changing population density (Table 1.2). This phenomenon was discovered and studied in detail by Boris Uvarov (Uvarov, 1938), later further investigations were added by Boris himself (Uvarov, 1977) and Meir Paul Pener (Pener, 1991; Pener and Yerushalmi, 1998; Pener and Simpson, 2009). It is now well reorganized that phase changes play an important role in swarm formation and its mass migration (Buhl et al., 2006; Gray et al., 2009). Recently, much emphasis has been shifted to locus behavioral ecology. As overcrowding starts in locust population, the behavior of members shifts from solitary phase to gregarious phase. These behavioral changes result in individuals to aggregate rather than avoiding one another. This provides positive feedback for crowding and triggers the phenotypic changes and slowly leads to change in morphology and coloration (Simpson and Sword, 2009). Other factors which contribute in phase polyphenism are pathogen resistance and warning coloration. This gregarious phase continues until the environmental factors become adverse and the population declines. The migratory swamps may merge with other colonies which they may encounter during migration (Sultana and Saeed, 2015).

TABLE 1.2 Terminologies Used for Locust Groups.

Terminology	Description
Assemblage	When the prime concern is pest management and economic loss
Inventory	When the prime concern is conservation
Population	When there is a prime importance is ecological and quantitative changes
Community	When there is a prime importance is ecological and interactive changes like competition and mutualism
Guild	When there is a prime importance is ecological and the use of common resources need to be highlighted
Swarm	Used to identify collective behavior and migratory units

Source: Adapted from Lockwood (2011).

1.3 LIFE FORMS AND LIFE ZONES

The information on locust habit and habitat is quite inadequate and confusing as the information is collected through general impression, short-term collection, mostly targeting adults, and those ignoring oviposition. The following life forms are widely recognized:

a) Terricoles
b) Arboricoles
c) Herbicoles
d) Graminicoles
e) Aquaticoles

1.3.1 TERRICOLES

Terricole species live on ground and feed on herbs without climbing them, they lay eggs in the ground. They have short body, nearly twice as wide as it is wide; the face is slightly oblique or vertical, and metathorax is above one. Their habitat predominates with bare ground. They show remarkable camouflage due to color relationship with habitat. Wing size may be suppressed or completely lost in some subdivisions (Roessingh et al., 1993). They are grouped into the following subdivisions:

a) Deserticoles, living in open deserts
b) Arenicoles, living on sand
c) Saxicoles, living on unfastened rocks

1.3.2 ARBORICOLES

These are the species which live on dense herbs, shrubs, grasses, and trees; usually on foliar part, and are active climbers. Their presence on the forest floor is also reported. Their body is laterally compressed, having low width-to-height ratio. Presence of leaf-like tegmina are the main features.

1.3.3 HERBICOLES

These species live mostly in dense herbs and shrubs with or without admixture of grasses. Their body is less cylindrical, face weakly oblique, moderately elongated; prosternal processes are well developed. They may lay eggs on plants.

1.3.4 GRAMINICOLES

The habitat of these species is predominantly grassland. Their body is laterally compressed, more elongated; they have strong mandible muscles, long ensiform antennae on their head and their body color is mainly green or straw-yellow (corresponds to the shades of the grasses). They lay eggs on the ground, with some exceptions, ovipositors are very much appropriate for digging in the ground.

1.3.5 AQUATICOLES

They live on the flat surface of large floating leaves, they show many similarities in most characteristics with terricoles. Their integument is smooth, they have well-developed tarsal, enlarged spines;their general coloration is similar to plant species that live in water. They are adapted for marshy areas, can swim and climb on plants.

1.3.6 LIFE ZONES

The concept of life zones was established to group areas having somewhat same communities. The eco-fauna of any area is directly dependent on the local vegetation cover of that area (Despland et al., 2004).

1.3.6.1 TEMPERATE

This area has moderate temperature, not very hot or very cold. The most prevalent species in this zone are *Duroniella laticornis, Cataloipus cognatus, Eyprepocnemis alacris, Choroedocus illustris,* and *Dericorys tibialis.*

1.3.6.2 TUNDRA

The characteristic feature of tundra is the lack of trees and the presence of sedges and grasses. The species are little investigated, however, *Bohemanella frigid* is mostly associated with Siberia and similar areas.

1.3.6.3 TAIGA

The predominant vegetation in this area is the coniferous forest which is the largest terrestrial biomass. These forests are mostly damaged, cleared, or thinned over the year and a lot of non-native vegetation has invaded these areas.

1.3.6.4 DECIDUOUS FORESTS

This area is inhabited by herbicolous species, such as *Primnoa*, *Ognevia*, *Eirenephilus*, *Melanopline*, and graminicolous species, such as *Chrysochra* and *Chorthippus*.

1.3.6.5 MEADOWS

The characteristic vegetation of this area is mesophilous grasses with scattered woody species. A lot of locust and grasshopper species are present in this area except *Calliptamus, Oedipodia*, *Acrotylus.*

1.3.6.6 GRASSLAND

These areas may have thick and thin vegetations belonging to the family Poaceae. The dominant species of these areas are herbivorous and terricolous; important species include *Calliptamus barbrus, C. italicus, C. coelesyriensis, Notostaurus albicornis, Pyrgodeoa armata.*

1.4 FEEDING HABITS

According to the feeding behavior, locusts have the following groups (Uvarov, 1977):

a) Multivorous (polyphagous): those which feed on different plants.
b) Ambivorous.
c) Herbivorous: those who nourish on herbs.
d) Arbivorous: those who nourish on tree leaves or shrubs.
e) Graminivorous: those who nourish on grasses.
f) Univorous: those which feed on single host plant.
g) Carnivorous: those which feed on other grasshopper species.

According to food selection, the following categories are reported Otte and Joern (1977)

a) Monophagous: their food selection is based on only one genus of the plant.

b) Oligophagous: their food selection is limited, usually within one plant family.
c) Polyphagous: their food selection is wide, they feed on a number of different plant families.

1.5 HISTORICAL OUTBREAK

Large desert locust outbreaks are reported in the literature over the years, these outbreaks are directly linked with warm and dry climates (Meynard et al., 2017). Recently, heavy rain and hurricanes resulted in climate change, such as high temperature and set drought conditions which are suitable for the locust outbreak in Arabian Peninsula. As per the FAO (Food and Agriculture Organization) record, the year 2020 outbreak was one of the deadliest in 70 years, which destroyed 70,000 ha of land in Ethiopia and Somalia (FAO, 2020). One of the most prominent locust outbreaks was in 1980, which occurred in Africa and has affected around 360,000 ha of crops in Northern Senegal (Walsh, 1986). In Australia, the year 1998 was the year of the locust outbreak where locusts attacked grasslands (Hinton and Library, 2007). Successively, in 2010, another locust spell occurred in New South Wales and it destroyed 390,000 km^2 area, this counts for about 49% of the cultivated area, and losses of around US$2 billion $ loss have been reported (Miao et al., 2015). Southern Russia (Stavropol region) and yellow river (China) were under locusts' attack in 2001, damaged 267,000 ha of arable grassland and 100,000 individuals, respectively (Zhu, 2001). The second wave of locusts emerged in between 2003–2004 in China's grasslands, where around 7 billion kg of crops were damaged (Marei et al., 2015).

1.6 RECENT OUTBREAK

Because of a worldwide increase in climate change phenomenon and rise in temperature, in 2020, the number of locusts could possibly increase by 64 million times; this could terribly flare-up the locust outbreak compared to last 100 years (Sharmila, 2020). With the beginning of the monsoon season in March, the locust outbreak is anticipated to further grow, this may result in food shortage and humanitarian emergencies (Ali et al., 2021). The previous patterns propose that locust outbreak events have a direct correlation with climate change, sea level rise, topography, latitude, temperature, and rainfall (Zhang and Liu, 2018). Climate change prompts more precipitation and higher temperatures, which speed up the rapid development and propagation of locusts (Gall et al., 2019). Under high temperature and humidity conditions, the recurrence of locust increases, and at temperatures around 40°C and humidity of 60–70%, the locust migrates and attacks new areas (Veran et al., 2015). The western areas of Africa and Australia are experiencing more fluctuations in temperature, moisture, and CO_2, which may result in more locust outbreaks in these areas (Li et al., 2018).

The historical records of locust outbreak point out the linkage with two important drivers, such as anthropogenic climate change and earth temperature rise, these two factors are the direct result of economic and political conflicts (Salih et al., 2020). The new locust event that began in Africa in February 2020 is likely due to a temperature rise combined with 210 days of bushfire in Australia and a heavy rainy season in the Middle East (FAO, 2020). It rose out of East Africa, moving to Southwest Asia and the Red Sea, causing a huge plague in Somalia, Ethiopia, and Kenya with concurring food security. Locust generations synchronize with autumn and summer giving brilliant premise to the huge outbreak (Zhang and Liu, 2018). Later, in Kenya, enormous locust outbreak emerged which measured 60 km long and 40 km wide; they attacked the northern Kenya in a month, making critical damage to crops, domesticated animals, and fields (Sharmila, 2020). This is the most terrible locust outbreak in the last 25 years in Somalia, Ethiopia, and Kenya. The swarms then moved toward Guangxi and Yunnan (China). In Yunnan, the major crops were potatoes, peanuts, tea trees, corn, and rice, which were damaged, and huge financial losses were encountered. Even a small locust group may eat much food in one day as about 2500 people or 10 elephants. (Hydrick et al., 2020). The food security and ecological balance in Africa are suspected to be altered or influenced by the locust outbreak in 2020. Indeed, around 1,000,000 insect species are recognized around the world, representing about 60% of the world's biodiversity (Jongema, 2015). These insects, at present, are used as protein source and meet the dietary needs (Arnold et al., 2007). With the increase in human population and diminishing food quality, meat production is decreasing, and this may be overcome by the utilization of insects as food. Just 2 kg of plant biomass or waste is required to produce 1 kg of insect biomass which is just 20% of meat production and can turn the economy positive (Khusro et al., 2012).

1.7 CONTROL MEASURES

Sustainable control measure is the need of the hour to minimize the use of toxic chemical pesticides and use of more eco-friendly technologies for the long-term betterment of ecosystem (He and Du, 2013). Malathion was mostly used in 1980 to kill locusts on 360,000 ha in Senegal, but in 2020, the world is looking for more sustainable ways to control locusts. One of the alternative ways is the use of microorganism s (like bacteria and fungi) as biological control agent against locust pest (Dakhel et al., 2019). Another way is to use the predators of locust larva, such as spider, mite, birds. This may reduce the locust population up to 90%. Crop rotation is also an emerging way to control the locusts' attack, keeping the crop sowing and harvest time mismatched, then locust hatching will reduce the disaster (Arroyo et al., 2019). Modification of landscape and converting lowland and marshy areas into fish ponds can reduce the possible locust breeding sites (Wang, 2011). Mechanical and physical traps can also serve as good options; these can be effective to control locust outbreak, and their efficiency can be merged with light and sound attractants. These lights have special wavelengths and specific sound wavelengths to attract insects toward the trap. The trapped locusts may later be killed or fed to chickens (Maleki and Khorram, 2010). Remote sensing is also gaining popularity in order to monitor the locust breeding sites, developing areas, and migration routes, but this approach is limited in use only for large-scale outbreak (Antoaneta, 2020).

1.8 PREVENTIVE MANAGEMENT

The locusts have been the greatest challenges to agriculture from the start of civilization (Katel, et al., 2021). The locusts are considered one of the most destructive and deadliest pests (Joshi et al., 2020; Lazar, et al., 2016). According to the FAO, the plague epidemic represents an "unusual risk" in areas that are already facing challenges due to climate change for livelihoods and food security (FAO, 2020; Banik, et al., 2020). Locusts are a serious threat to food security. Locusts threaten food security because of the destruction of crops, which reduces food supplies for humans. In addition, locusts devour grass and natural wild plants which affect animals and wildlife as well (Escorihuel, et al., 2018). The world's one-tenth of the population's livelihoods are expected to be affected by locust (FAO, 2021).

Different monitoring campaign costing many dollars has been carried out. Unfortunately, there was no great outcome despite serious environmental issues due to several chemical insecticides used for locust control (Katel, et al., 2021). It is crucial to identify locust populations early to avoid long and expensive procedures before reaching plague levels (Ibrahim, 2008). Locust plague control can be very expensive economically.

1.8.1 STAGES OF PREVENTIVE MANAGEMENT

To prevent means to avert the condition or event by preventing its occurrence. In the case of locust, preventive management can occur in three stages.

1.8.2 UPSURGE PREVENTION

This involves the measures made right before the attack by locusts. This method focuses on preventive measures for the protection of agricultural cropland. The interventions are taken following the onset and plague of desert locusts which are economically and socially dangerous due to a previous lack of preparation for response. This prevention aims to prevent croplands from destruction rather than attacking primary breeding zones (commonly in remote regions). Unfortunately, major upsurges and epidemics are linked to severe economic and social disturbances (Sharma, 2014).

1.8.3 OUTBREAK PREVENTION

Outbreak prevention is the management of gregarization; it involves interventions before or during the change from solitary to gregarine phases of pest attack (Showler, 2019). The prevention of outbreaks is an early enough action to avoid the onset and gregarization initial development before or when the locust nymphs are in small parts and adults are in loose groups. This technique will reduce enough population to decrease the crowd-related pheromone indicators (the crowd generates positive feedback for the growth of entirely gregarious populations), causing overcrowding. Theoretically, the sustainable intervention with the triggers of semi-chemical gregarization can sustain the recession stage permanently (Meynard, et al., 2020). Although proaction is better than reaction, a preventive strategy offers a rather more effective control choice early.

1.8.4 UPSURGE ELIMINATION

Epidemic intervention is correctly defined as "proaction" to stop a spread of an outbreak toward the status of plague. Proaction is an intermediary prevention between upsurge and outbreaks prevention. The expansion of locust populations from recession to status of plague is continuous; the outbreak prevention occurs at an early stage and is followed by proaction after intervention in outbreaks. While these two words relate to early actions to prevent epidemics, they differ in terms of the timeframe (Cressman, 2020). The key difference between the two is obscured by the outbreak prevention following proaction. Historical data demonstrate that the greater the time of gregarious populations breeding, the more likely they are to get overwhelmed and the greater the possibility that they would become plague (Pandey, et al., 2021).

1.8.5 STRATEGIES FOR PREVENTIVE MANAGEMENT

The strategies used for preventive management are the most reasonable, practical, economically viable, and environmentally sound ways for controlling locust control, and are increasingly applied in preventive management techniques. Strategic plans are required to successfully manage the locust outbreak, including financial coordination, integrated management, labor division, international assistance, and effective use of resources (Zhang, et al., 2019). The worldwide community has agreed that proactive approaches or, if feasible, prevention is preferable to uncontrolled locust population, occasionally leading to upheavals and diseases (Belayneh, 2005). Preventive management is intended to treat as many hot spots as possible before crops damage by early intervention in outbreaks. Some of the best preventive management methods have been with species whose early population expansion is mostly limited to some areas which are breeding-friendly sites. The control and treatment of populations within the outbreak zones could prevent plagues from high population rises and massive escape in crop areas in the future (Pioua, et al., 2013). Because of the large-scale impact and cost of control, monitoring is crucial for locust management. The preventive management of locust is a complex process which requires multidisciplinary approaches, including ecology, biology, entomology, modeling of the spatial distribution, weather predictions, climate analyses, and organisms and interactions between them (for example, sheep and birds), bioagents or chemical insecticides control, and remote sensing applications (Klein, et al., 2021). In technologically challenging locations, it is not possible to monitor the breeding of locusts, so efficient prevention activity is impossible to carry out. Managing tiny hopper bands and swarms before increasing size and numbers to form an outbreak is the best preventive method (Waldner, et al., 2015). Two actions are taken for the effective management of locust control: monitoring/surveillance and elimination. Monitoring is important for the identification of populations to be eradicated.

1.8.6 MONITORING

For prevention and proaction to be sustainable, it requires improved technologies to monitor and control. Monitoring incorporates numerous elements and technological levels, whether in recessions or epidemics. Farmers' and nomads' reports require comparatively little technological sophistication, but government surveys usually involve sensors, cars, radios, and training (Cressman, 2008). Since the mid-1990s, technological improvements include remote sensor images and climate and meteorological data collection, precisely predicting locust activities. In addition, e-mail and internet communication allow the speedy reporting and transmission of the information collected (e.g., locust, weather, images of vegetation) from nations affected by desert locusts (Lecoq, 2001). The success of interventions in 2007–2016 indicates that these instruments helped to suppress the population efficiently.

1.8.7 REMOTE SENSING

Remote-sensor-based research applications and case studies were major drivers for understanding ecological and environmental factors significant to the locusts. Since the 1980s, remoteness data gathered information has been used to manage numerous locusts' attacks and has helped enhance and more effective control of locusts and plagues worldwide. The Food and Agriculture Organization (FAO) has effectively introduced standardized monitoring procedures and data gathering when remote sensing data and applications play a crucial role (Hielkema, 1990). The remote sensing data for locust outbreaks were initially introduced by Pedgley (1997) and Hielkema (1990) and later implemented in the locust information service of FAO (DLIS). Hielkema (1990) and Hielkema and Snijders (1994) concentrated on the cloud imaging of Meteosat for estimating rainfall and on the vegetation development estimate for Landsat and AVHRR. The targets achieved using remote sensing are the following.

- Mapping and monitoring the state and environmental factors of the locusts encourage the process of transition between solitary and gregarious phases. Historical data, current vegetation monitoring, and forecasts predict locust hatching time and possible epidemics.
- Airborne or UAV-based sensors for locust nympho bands and swarm monitoring.
- Evaluation of vegetable and crop damage post-outbreak.

Desert locusts are difficult to control, but the strategies necessary for the control include proactive monitoring, initial intervention, and focused implementation (Samejo, et al., 2021). Increased knowledge of pest biology and ecology and more effective monitoring and control methods are now the key elements in the preventive management. The efforts for locust control continue to rely on chemicals.. The monitoring and forecasting of populations are the two aspects of management of locusts (Zhang, et al., 2019).

1.9 GREGARIZATION RISK AS A PART OF PREVENTIVE LOCUST MANAGEMENT STRATEGIES

To reduce the potential gregarization risk and keeping check on the locust population as management strategies, here are three developments. The first development is the temporal and spatial scale monitoring of desert locust, potential area where conditions are favorable (van Huis et al., 2008). Many authors have reported the monitoring of localized areas in time and space, merged with weather conditions. Around 0.8% area is suitable for gregarization in Africa and Latin America. The second development is the preventive management strategy, this must be customized on regional basis. This needs historical records, weather forecast, metrological records, remote sensing data, and ground surveillance teams. The major limitation in ground surveillance teams is the lack of expertise and entomological knowledge. Beside other resources, trained or skilled human resource is the critical factor for the success of any locust control program. The third key development is the existence of quantitative predictive models to predict the future trends and risks. These predictive models need local data and other parameters (resource availability and abundance) for accurate calculation, prediction, and modeling (Pener and Simpson, 2009). Pener and Simpson (2009) developed a concept frame work to synchronization of gregarization risk and locust control. This concept frame work was developed to utilize all the available resource and data to make better strategies to control locust. In many countries, the government agencies visit and monitor potential breeding sites, the frequency of visit depends on the low and high risk of gregarization. The higher the risk, the more frequent should be the monitoring, this successively generate more accurate migration and risk models. The strict protocols or standard operating principles should be developed and implemented for surveys. In the presence of protocols, the cost reduces, the efficiency of the team improves, and there is also an improvement in data accuracy, time management, but the resource need decreases.

1.10 LUBILOSA

LUBILOSA is a French name for a research program which was started to develop the biological control of locust as an alternative method to chemical control. This was a 13-year long project, and it aimed to develop a commercial biopesticide for locust. This program successfully isolated an entomopathogenic fungus which is virulent to locust. This research project was developed by David Greathead and Chris Prior based at the UK International Institute of Biological Control (IIBC). LUBILOSA collaborated with GTZ's locust program. Coolaboration was also established with the crop protection agencies of Gambia, Senegal, Mali, Chad, Burkina Faso, Benin and Niger. The financial grant of US$17,000,000 was contributed by the following donors:

- United States Agency for International Development (USAID)
- Swiss Agency for Development and Cooperation (SDC)

- Dutch Directorate General of International Cooperation (DGIS)
- Department for International Development, UK (DFID)
- Canadian International Development Agency (CIDA)

The whole program consisted of four phases. During the first phase (1990–1992), different biological options were examined and tested, and certain species belonging to the phylum Deuteromycota were isolated (about 180 species). These species have lipophilic spores and have a tendency to grow rapidly on artificial media. The idea at this stage was conceived to develop ultra-low volume (ULV) oil-based formulation of fungal spores. This was to maximize the efficacy and to use the same equipment which was previously in operation. The technical feasibility of the isolates was tested in the laboratory and *Metarhizium acridum* was selected as the most effective among the isolated ones. During the second phase (1993–1995), field trials were conducted to test several oil formulations. These trials were successful. The major challenging factor during field trial was the high mobility of locust swamps. The research team did not opt the idea of fixed plot trials, even if the locusts spread over several square kilometers in size. The human and technical capacities were enhanced in spore production, spore separation, cross-contamination control, spore drying, and lastly packaging. In the third phase (1996–1998), the locust behavior, ecotoxicology, and mass production and commercialization were studied. The field trials were setup against desert locusts, brown locusts, and Sahelian tree locusts. In the field trials, it was observed that *Metarhizium* reduced the locust population up to 90% in 3 weeks. One prominent success compared with that of chemical pesticide was that in the presence of chemical pesticide, the locust population starts appearing again within 2 weeks. However, biological control limits the reappearance of locust to one and a half months. Spores of *Metarhizium* may survive unfavorable conditions and between two seasons. One important finding durind field trials was that the locust species were able to detect that they are infected and thus they change their ecological behavior. They started to spend more time under sun and increase their body temperature up to 40°C. Another target of the third phase was to study the environmental impact of the selected *Metarhizium* isolate. Different laboratories carried out the ecotoxicology test in different ecological zones, the spores were non-harmful to birds, mammals, fish, and reptiles and do not infect other higher species. Some species of silk worms, parasitoid wasps, honey bees, and termites may get mild infection under laboratory conditions. The mass production was done by firstly by using liquid fermentation and later followed by a solid substrate phase. A novel machine "Mycoharvester" was invented to improve the efficiency and to make the process more robust. This machine passed the quality standards and was presented for commercial license. The machine was also optimized for viable spore count, target pests, and particle size spectrum. For commercialization, two companies, such as "Biological Control Products" and "Natural Plant Protection" were involved. These companies were already involved in biopesticide business in African countries. Biological Control Products got the registration and license under the name "Green muscle." Natural plant protection later halted their operations and "Biological control protection" was taken over by Becker Underwood in 2001.

The fourth stage of this program was mostly designed for the promotional activities. Convincing the governmental and intra-governmental agencies and penetrating the market were the challenging tasks, especially in the start. Later, large pesticide selling companies got interested and offered all support and financial solution for those procuring their products. Food and Agriculture Organization of the UN (FAO) became interested at the end of this program and these serious efforts continued to market and the use of "green muscle." The post-LUBILOSA was also very challenging, however, under FAO support, the research, field trials, optimization, and commercialization continued to increase. At one stage, the mixture of lambda-cyhalothrin and Green Muscle was developed to improve the efficacy. In 2019, a Switzerland-based company "Elephant Vert" developed a new product named "NOVACRID" using *M. acridum,* and right now it has sole rights for production and sales.

KEYWORDS

- **storm**
- **climate change**
- **food security**
- **hunger**
- **poverty**

REFERENCES

Antoaneta, R. Why Gigantic Locust Swarms Are Challenging Governments and Researchers. *Nature* **2020,** *579* (7799), 330.

Arnold, V. H.; Keith, C.; Joyce, I. M. Preventing Desert Locust Plagues: Optimizing Management Interventions. *Entomol. Exp. Appl.* **2007,** *122* (3), 191–214.

Arroyo, A. S.; Martinez, G. J.; Pereda, S. M.; Sierra, F. D. First Breeding Record of Grasshopper Sparrow in Northern Durango, Mexico. *Huitzil. Revista Mexicana de Ornitología* **2019,** *20*, 1–7.

Ali, A.; Nasir, A.; Muaaz, M. Evaluation of Monsoon Effect on Groundwater Quality Using GIS Mapping: A Case Study of Bahawalnagar District - Pakistan. *J. Qual. Assurance Agric. Sci* **2021,** *1* (1), 30–36.

Banik, A. et al. Screening and Potent Applicability Analysis of Commonly Used Pesticides Against Desert Locust: An Integrative Entomo-Informatics Approach. *bioRxiv*, 2020.

Belayneh, Y. T. Acridid Pest Management in the Developing World: A Challenge to the Rural Population, a Dilemma to the International Community. *J. Orthoptera Res.* **2005,** *14* (2), 187–195.

Branson, D. H. Effects of Altered Seasonality of Precipitation on Grass Production and Grasshopper Performance in a Northern Mixed Prairie. *Environ. Entomol.* **2017,** *46* (3), 589–594.

Buhl, J.; Sumpter, D. J. T.; Couzin, I. D.; Hale, J. J.; Despland, E.; Miller, E. R.; Simpson, S. J. From Disorder to Order in Marching Locusts. *Science* 2006, *312*, 1402–1406

Cressman, K. The Use of New Technologies in Desert Locust Early Warning. *Outlooks Pest Manage.* **2008,** *19* (2). s.l.:Research Information.

Cressman, K. Technical Innovations in Global Early Warning in Support of Desert Locust Area-Wide Management. In: *Area Wide Integrated Pest Management.* s.l.:CRC Press, **2020;** p. 9.

Crook, D. R.; Robinson, B. E.; Li, P. The Impact of Snowstorms, Droughts and Locust Outbreaks on Livestock Production in Inner Mongolia: Anticipation and Adaptation to Environmental Shocks. *Ecol. Econ.* **2020,** 177.

Dakhel, W. H.; Latchininsky, A. V.; Jaronski, S. T. Efficacy of Two Entomopathogenic Fungi Metarhizium Brunneum Strain F52 Alone and Combined with Paranosema Locustae Against the Migratory Grasshopper Melanoplus Sanguinipes Under Laboratory and Greenhouse Conditions. *Insects* **2019,** *10* (4), 94.

Despland, E.; Rosenberg, J.; Simpson, S. J. Landscape Structure and Locust Swarming: A Satellite's Eye View. *Ecography* **2004,** *27*, 381–391.

Escorihuel, M. J. et al. SMOS Based High Resolution Soil Moisture Estimates for Desert Locust Preventive Management. *Remote Sens. App* 2018, *11*, 140–150.

FAO *Five Things You Should Know About an Age-Old Pest: The Desert Locust,* s.l.: FAO, 2020.

FAO *Locust Watch,* s.l.: FAO, 2021.

FAO, 2020. http://www.fao.org/home/common-elements/top-navigation-content/about/en/.

Fontana, P.; Marino-p ˜ ´erez, R.; Sanabria-urb´ an, S.; Woller, D. A. Studies in Mexican Grasshoppers: Three New Species of Dactylotini (Acrididae: melanoplinae) from

Mexico and a Review of Existing Conspecifics with Comments on Their Geographical Distributions. *Zootaxa* **2017,** *4337* (3), 301–343.

Gall, M. L.; Overson, R.; Cease, A. A Global Review on Locusts (Orthoptera: Acrididae) and Their Interactions with Livestock Grazing Practices. *Front. Ecol. Evol.* **2019,** *7*, 263.

Gray, L. J.; Sword, G. A.; Anstey, M. L.; Clissold, F. J.; Simpson, S. J. Behavioural Phase Polyphenism in the Australian Plague Locust (Chortoicetes terminifera). *Biol. Lett.* **2009,** *5*, 306–309.

He, M. J.; Du, P. Field Efficacy Test of 2% Thiametholine Microcapsule Suspension for Locust Control. *Rur. Sci. Technol.* **2013,** 4, 38–39.

Hielkema, J. U. Satellite Environmental Monitoring for Migrant Pest Forecasting by FAO: The ARTEMIS System. *Biol. Sci* 1990.

Hinton, B.; Library, P. Plague Locust Threat to Australian Agriculture. *Euphytica* **2007,** *154* (3), 289–294.

Hydrick, A.; Masys, A.; Izurieta, R.; Reina, O. M. Agricultural Emergencies: Factors and Impacts in the Spread of Transboundary Diseases in, and Adjacent to, Agriculture. *Global Health Sec.* **2020,** 13–31.

Ibrahim, K. M. Plague Dynamics and Population Genetics of the Desert Locust: Can Turnover During Recession Maintain Population Genetic Structure? *Mol. Ecol* **2008,** *10* (3), 581–591.

Hielkema, J. U.; Snijders, F. L. Operational Use of Environmental Satellite Remote Sensing and Satellite Communications Technology for Global Food Security and Locust Control by FAO: The ARTEMIS and DIANA Systems. *Acta Astronautica* **1994,** 32 (9), 601–616.

Jongema, E. *World List of Edible Insects*; Wageningen University: Wageningen, The Netherlands.

Joshi, M.; Varadharasu, P.; Solanki, C.; Birari, V. Desert Locust (Schistocera gregaria F.) outbreak in Gujarat (India). *Agric. Food: E-Newslett.* **2020,** 2 (6), 691–693.

Katel, S. et al. Desert Locust (*Schistocerca gregaria* Forskal) and Its Management: A Review. *J. Agric. Appl. Biol.* **2021,** 2 (1), 61–69.

Khusro, M.; Andrew, N. R.; Nicholas, A. Insects as Poultry Feed: A Scoping Study for Poultry Production Systems in Australia. *World Poult. Sci. J.* **2012,** *68* (3), 435–446.

Klein, I.; Oppelt, N.; Kuenzer, C. Application of Remote Sensing Data for Locust Research and. *Insects* **2021,** *12* (3), p. 233.

Lazar, M.; Piou, C.; Mitiche, B. D.; Lecoq, M. Importance of Solitarious Desert Locust Population Dynamics: Lessons from Historical Survey Data in Algeria. *Entomologia experimentalis et applicata* **2016,** *161* (3), 168–180.

Lecoq, M. Recent Progress in Desert and Migratory Locust Management in Africa: Are Preventative Actions Possible? *J. Orthoptera Res.* **2001,** *10*, 227–291.

Li, R. R. Causes of Grassland Locust Disaster and Control Measures. *China Livestock Poultry Breed.* **2018,** *14* (6), 27.

Lockwood, J. A. *The Ontology of Biological Groups: Do Grasshoppers Form Assemblages, Communities, Guilds, Populations, or Something Else?* Hindawi Publishing Corporation, Psyche, **2011**; p 9.

Maleki, M.; Khorram, M. S. Bayesian Network and Pest Management: A Case Study of Trap Plants on Locust Population. *Adv. Environ. Biol.* **2010,** *4* (2), 147–151.

Marei, N. H.; Samiee, E. A. E.; Salah, T.; Saad, G. R.; Elwahy, A. H. M. Isolation and Characterization of Chitosan from Different Local Insects in Egypt. *Int. J. Biol. Macromol.* **2015,** *82*, 871–877.

Meynard, C. N.; Lecoq, M.; Chapuis, M.-P.; Piou, C. On the Relative Role of Climate Change and Management in the Current Desert Locust Outbreak in East Africa. *Global Change Biol.* **2020,** 26 (7), 3753–3755.

Meynard, C. N.; Gay, P. E.; Lecoq, M.; Foucart, A.; Piou, C.; Chapuis, M. P. Climatedriven Geographic Distribution of the Desert Locust During Recession Periods: Subspecies' Niche Differentiation and Relative Risks Under Scenarios of Climate Change. *Global Change Biol.* **2017,** *23* (11), 4739–4749.

Miao, J.; Zhao, Z.; Liu, H. L. Using MODIS Data to Monitor the Change of Vegetation Index of Diseases and Insect Pests in a Wide Range—Taking the Locust Disaster in Australia in 2010 as an Example. *China Agron. Bull. (Arch. Am. Art)* **2015,** *31* (14), 148–155.

Pandey, M. et al. Desert Locust Invasion in Nepal and Possible Management Strategies: A Review. *J. Agric. Food Res.* **2021,** *5*, 100166.

Pedgley, D. ERTS Surveys a 500 Km2 Locust Breeding Site in Saudi Arabia in Proceedings of the Third Earth Resources. In *Third Earth Resources Technology Satellite-1 Symposium.* s.l.:s.n., **1997;** pp 233–246.

Pener, M. P. Locust Phase Polymorphism and Its Endocrine Relations. *Adv. Insect Physiol.* **1991,** *23*, 1–79.

Pener, M. P.; Simpson, S. J. Locust Phase Polyphenism: An Update. *Adv. Insect Physiol.* **2009,** *36*, 1–286.

Pener, M. P.; Yerushalmi, Y. The Physiology of Locust Phase Polymorphism: An Update. *J. Insect Physiol.* **1998,** *44*, 365–377.

Pener, M. P.; Simpson, S. J. Locust Phase Polyphenism: An Update. Adv. Insect Physiol. **2009,** *36* (1st ed.). Academic Press; p *9.* ISBN 9780123814289.

Pioua, C. et al. Coupling Historical Prospection Data and a Remotely-Sensed Vegetation Index for the Preventative Control of Desert Locusts. *Basic Appl. Ecol.* **2013,** *14* (7), 593–604.

Roessingh, P.; Simpson, S. J.; James, S. Analysis of Phase-Related Changes in Behaviour of Desert Locust Nymphs. *Proc. R Soc. B* **1993,** *252*, 43–49.

Salih, A. A. M.; Baraibar, M.; Mwangi, K. K.; Artan, G. Climate Change and Locust Outbreak in East Africa. *Nat. Clim. Change* **2020,** *10*, 584–585.

Samejo, A. A.; Sultana, R.; Kumar, S.; Soomro, S. Could Entomophagy Be an Effective Mitigation Measure in Desert Locust Management? *Agronomy* **2021,** *11* (3), 455.

Sharma, A. Locust Control Management: Moving from Traditional to New. *Entomol. Ornithol. Herpetol.* **2014,** *4* (1).

Sharmila, D. Locust Swarms in East Africa Could Be a Catastrophe. *Lancet* **2020**. *395* (10224), 547.

Showler, A. T. Desert Locust Control: The Effectiveness of Proactive Interventions and the Goal of Outbreak Prevention. *Am. Entomol.* **2019,** *65* (3), 180–191.

Simpson, S. J.; Sword, G. A. Phase Polyphenism in Locusts: Mechanisms, Population Consequences, Adaptive Significance and Evolution. In *Phenotypic Plasticity of Insects: Mechanisms and Consequences. Science Publishers*; Whitman, D., Ananthakrishnan, T. N., Eds.; Einfield, **2009;** pp 145–189.

Steedman, A., Ed., Locust Handbook, 2nd ed.; Natural Resources Institute: London, **1988**. ISBN 978-0859542326.

Sultana, R.; Saeed, M. *Grasshoppers and Locusts of Pakistan*; Higher Education Commission: Pakistan, **2015**. ISBN: 978-969-417-180-7

Uvarov, B. *Grasshoppers and Locusts*, vol. 2; Centre for Overseas Pest Research: London, 1977.

Uvarov, B. P. Locust as a World Problem. In Premie` re confe´ rence internationale pour la protection contre les calamite´ s naturelles; Paris, Sept 13–17. Commission franc¸aise d'e´tudes des calamite's with the Support of Union Internationale de Secours, pp 376–382.

Veran, S.; Simpson, S. J.; Sword, G. A.; Deveson, E.; Piry, S.; Hines, J. E.; et al. Modeling Spatiotemporal Dynamics of Outbreaking Species: Influence of Environment and Migration in a Locust. *Ecology* **2015,** *96* (3), 737–748.

Waldner, F. W.; Ebbe, M. A. B.; Cressman, K.; Defourny, P. Operational Monitoring of the Desert Locust Habitat with Earth Observation: An Assessment. *ISPRS Int. J. Geo-Inf.* **2015,** *4* (4), 2379–2400.

Walsh, J. Return of the Locust: A Cloud over Africa: Last Year's Rains Brought Better Harvests–and a Bigger Crop of Locusts and, Grasshoppers, Threatening a Revival of the Plagues of the Past. *Science* **1986,** *234* (4772), 17–19.

Wang, J. W. The Main Causes of Locust Disaster and Its Comprehensive Control Technology. *Pest. Market Inf.* **2011,** *43*.

Wikipedia Contributors. List of Locust Species. In *Wikipedia, the Free Encyclopedia*, July 28, 2020. https://en.wikipedia.org/w/index.php?title=List_of_locust_species&oldid=969923542 (accessed June 15, 2021).

Yang, X.; Zhang, K.; Wang, J.; Jia, H.; Duan, J. Assessment of Genetic Diversity and Chemical Composition Among Seven Black Locust Populations from Northern China. *Biochem. Systemat. Ecol.* **2020,** *90*, 104010.

Zhang, J. T.; Liu, X. Y. Locust Disaster in the Central Plains of Yin and Shang Dynasties. *His. Natural Sci. Res.* **2018,** *37* (4), 417–423.

Zhang, L.; Lecoq, M.; Latchininsky, A.; Hunter, D. Locust and Grasshopper Management. *Annu. Rev. Entomol.* **2019,** *64*, 15–34.

Zhen, R. W.; Xia, Y. X.; Keyhani, N. O. Sex-Specific Variation in the Antennal Proteome of the Migratory Locust. *J. Proteomics*. **2020,** *216*, 103681.

Zhu, E. L. Dynamics of Locust Occurrence and Control in Foreign Countries. *Hubei Plant Protect* **2001,** *4*, 23–25.

CHAPTER 2

Effect of Locusts on Crops During the Last Decade in Agricultural History

WAJIHA ANUM[1], MADIHA MOBEEN KHAN[1], LAILA SHAHZAD[2], NAEEM ARSHAD MAAN[1], and IMRAN AKHTER[1]

[1]*Regional Agricultural Research Institute, Bahawalpur, Pakistan*

[2]*Sustainable Study Center, GC University, Lahore, Pakistan*

ABSTRACT

Agriculture sector has faced plagues of desert locusts (*Schistocerca gregaria*) across many areas of Asia, Africa, and Middle East. Their swarms have ability to fly up to large distance making them a potential insect for economically important crops, such as wheat, maize, sorghum, and others. Locust invasion is a severe threat to food security directly by destroying the livelihoods of the entire farming community as well as other associated sectors. Semi-arid areas are more prone to locust attack, thus, these regions face a "double" threats in the form of lesser rains, extreme temperature fluctuations. Insect pest attacks under such conditions aggravate the situation. Under such circumstances, there is a dire need to understand the complete ecological setup of the desert locust for ensuring accurate and timely control measures to be adopted especially in the case of plagues. This chapter will solely elaborate the complete ecological and morphological aspects of the insect and a special emphasis is made on the attack mechanisms, crisis situations, control mechanisms, and the effects on crops during the last decade.

Locust Outbreaks: Management and the World Economy. Umair Riaz, Khalid Rehman Hakeem, (Eds.)

2.1 INTRODUCTION

The desert locusts (*Schistocerca gregaria*) (Forskål, 1775) have always been a great menace for rural areas in terms of communal solidarity and lesser food security, where the sole earning comes from the cultivation of crops. Those areas with climatic conditions suitable for their growth and reproduction are more prone to damage. Arid and semi-arid areas with minimum rainfall already undergo water shortage and crops flourish at risk. In addition, the desert locusts' attack results in socioeconomic and environmental disturbances over prolonged time. The main problem creating characteristic of *S. gregaria* is their ability to migrate over larger distances, thus targeting far areas, and under plague conditions, it can damage whole crop fields across the globe (Lecoq, 2003).

Desert locust thrives in dry areas, such as deserts and semi-arid regions representing a main threat to a vast region ranging from Atlantic Ocean and North Africa to Middle East and Southwest Asia. *Schistocerca gregaria* inflicts detrimental effects on agro and animal husbandry along with pastors produce throughout its incursion times, it can disturb socioeconomic as well as environmental setup of an area. Being capable of highly developed migratory capacity, it can travel over large distances thus making huge damage at the international level. The outbreaks and upsurges develop from time to time and invasions occur, and this cycle is also linked to favorable rainfall occurrence. Only the interruption occurs during a phase commonly called recession period, the desert locust' solidarity population occurs in small amount and restricted to areas like deserts or areas that are far from cultivated lands (referred to as recession area). The recession area particularly concerns the Saharan zones and covers over 16 million km^2 of area, whereas during the invasion phase, the gregarious populations can invade 29 million km^2 area and mostly covers an extensive area of 65 countries in Africa, Southwest Asia, and Middle East. As this invasion area is mostly the cultivated land and popular one, so the damage affects beyond 1 billion people (Lecoq, 2003).

2.2 LOSSES TO CROPS

Under low population densities, locusts are solitary insects (they are in "solitarious phase") that cause little harm to agriculture. When crowded, locusts enter a "gregarious phase," forming dense and destructive swarms capable of marching long distances from their usual breeding areas and posing a serious threat to agriculture (Pener and Simpson, 2009).

The ample rains help in reproduction of Desert Locusts thus extending the epidemic. Such a situation poses a serious risk during peak crop-producing season. Desert locusts have achieved the status of most harmful pest species of migrant nature owing to their devastating damage to crops and ability to travel over longer distances (Cressman, 2016).

A typical swarm may consist about 150 million DL/km^2 having the ability to travel up to 150 km/day. A minute DL swarm of 1 km^2 will yet devor the same quantity of sustenance as 35,000 people in a single day (Food and Agricultural Organization (FAO, 2020e).

As a result, a swarm of desert locusts of 40–80 million locusts in less than 1 km^2 will consume roughly 190 million kg of plants per day. The desert locust plagues are estimated to impede one-tenth earth's population concerning their everyday maintenance. Devastating natural disasters are caused by locust plagues, wreak havoc on agriculture in many parts of the world, especially in Africa. For example, Morocco lost more than $50 million in one season due to desert locusts' attacks. Previously, Steedman (1988) stated a loss of 167,000 tons of grains in Ethiopia during locusts' attack in 1958, which was enough for feeding 1 million people for a year long.

During 2003–2005 upsurges, a traditional crop loss was estimated as 30% (Belayneh, 2005) without including the losses from small farms in Sahel, leading to extreme food insecurity (Showler, 2009). Furthermore, the upsurge resulted in farmer and pastoralist migration towards metropolitan regions, resource-centered struggles among growers, migrants, and pastoralists, a shortage of market supplies to balance farming product rates and global food relief intervention (FAO 2004a-e; FEWS 2004, Belayneh 2005, Doré et al. 2005). FAO (2006) reports a substantial loss in cereal production in Burkina (80%), Mali (90%), and Mauritania (100%). Similarly, legume crop loss was estimated to be 90% in the three countries. Mali and Burkina Faso faced a loss of 30% in pastures while Mauritania fodder production faced a loss of 85%. A crop loss of 10–20% (about $500 million) was also stated (Belayneh, 2005). According to the FAO (2006), 8.4 million people in six Sahelian countries were affected to varying degrees.

2.3 COVID AND LOCUST EFFECT

The situation in the globe has worsened with the combined effect of the three most devastating events, that is, floods, locusts, and COVID-19. It created a "Triple Menace" for East Africans (Global Agricultural

Monitoring (GAM), 2020). A similar situation has also been stated by Kassegn and Endris (2021) and the event has been given the name of "Triple Threat" (Kassegn and Endris, 2021).

According to the available data, almost 130 million people will be affected by hunger eventually by 2020 as a result of COVID-19's projected economic impact, generally in low-to-middle-income African countries which are prone to locust infestation (Rahaman et al., 2020). The increasing risks posed by COVID-19 and locusts infestation are amplified due to floods because they act as a barrier to the implementation of timely protective measures in place (IFRC, 2020).

2.4 SURVEILLANCE AND CONTROL

Other research studies suggest that there are other aspects of desert locust ecology's spatial–temporal aggregation dynamics that could be important to surveillance efforts. A field study in Mauritania stated that during their march, Desert locusts passby many trees and gradually rest on scattered trees before sunset, establishing aggregations (Maeno et al., 2018). The adults tend to rest on massive trees and larger bushes, however, gregarious bands of nymphs congregated on and around the local plant community's largest tree (Maeno et al., 2018).

For effective surveillance, it is important to get an advantage from studies that indicate variability in the defensive response of locusts shown toward any approaching thing including surveillance personnels or any predators. Fully grown locusts' defensive responses include instant flying, however, for flight, the minimum threshold temperature is $\approx$22°C (Maeno et al., 2018). At cooler temperatures, the locusts move to the middle of the bushes from the edges while those locusts individuals on taller plants ($\geq$2 m) tend to show no movement (Maeno et al., 2019). The hectares of land sprayed for locusts control are depicted in Table 2.1.

2.5 THE UPSURGE

During the winter and spring of 2020, the latest huge upsurge of DL was visible. The incident began when heavy rains occurred in Saudia Arabia (Rub al Khali, or Empty Quarter) in the second half of 2018 (FAO, 2018). The favorable breeding conditions that resulted helped three undetected generations (FAO, 2018), and swarms scattered across Saudi Arabia in

TABLE 2.1 Land Sprayed (Hectares) with Insecticides for Controlling DL According to Countries from 2010 to 2016 (FAO, 2017).

Country	2010	2011	2012	2013	2014	2015	2016
Algeria	1288	1244	53.342	17.095	84	0	1,417
Chad	0	0	2,630	0	0	0	0
Egypt	0	5288	9260	36621	82	0	0
Eritrea	0	920	0	59670	33311	17117	850
Ethiopia	0	0	0	0	6622	0	239
India	4700	0	0	0	0	0	0
Iran	0	6703	0	510	30100	0	0
Israel	0	0	0	28500	0	0	0
Kenya	0	0	0	0	0	0	0
Libya	40	0	21474	2755	0	0	0
Mali	850	1200	0	0	0	0	0
Mauritania	4402	59120	26053	22961	300	3915	23155
Morocco	1798	7039	2582	11643	0	17	15623
Niger	0	96	64557	805	541	0	0
Oman	0	0	0	0	5150	0	0
Pakistan	14204	8771	0	0	0	0	410
Palestine	0	0	0	16	0	0	0
Saudia Arabia	3526	86376	1363	94686	142481	10968	6420
Somalia	0	0	0	0	76	0	53
Sudan	14905	20643	34419	160289	148152	76145	5325
UAE	0	0	0	0	2500	0	0
Yemen	1450	0	0	46608	3832	0	614

December, as well as to Egypt, Sudan, Yemen, and Eritrea, and south-western Iran, where breeding took place (FAO, 2019a).

During whole spring (2019c), Saudia Arabia, Sudan, Egypt, Eritrea, and Iran undertook control mechanisms. However, Yemen was affected by violent strife, while poor control response occurred in Iran, the Iranian swarms infiltrated the India–Pakistan border and created three more generations of DL (FAO, 2020d). Yemeni swarms go across the Red Sea toward northern Somalia and Ethiopia where their breeding produced more swarms (FAO, 2018).

During spring 2019, there were no survey results from Somalia (FAO, 2019b); however, a possibility of swarm entrance from Yemen was reported and further they continued to reproduce. In Somalia and Yemen, where gregarious populations were not monitored, surveillance remained lax (FAO, 2019d). Desert locust activity resumed in Eritrea and Sudan in the fall and winter of 2019–2020, and began in Saudi Arabia, India, Pakistan, and Iran in December. In December 2019, Kenya was invaded, and infestations continued to spread through February 2020, followed by monitoring operations (FAO, 2020a). Furthermore, population of locust arriving in Oman from Saudi Arabia increased until interference in the second half of the winter. In Somalia and Yemen, extroverted populations continued despite poor surveillance and inadequate control (FAO, 2020b) operations. Infestations in Eritrea decreased during the spring of 2020 (March–June), while large gregarious populations and control operations persisted in Ethiopia, Kenya, Oman, Saudi Arabia, Iran, Pakistan, and by late spring in India (FAO, 2020c). In Somalia and Yemen, surveillance and control activities were inadequate, but gregarious populations were likely to be large (FAO, 2020c). Figure 2.1 represents the areas which were heavily infested during 2018–2021.

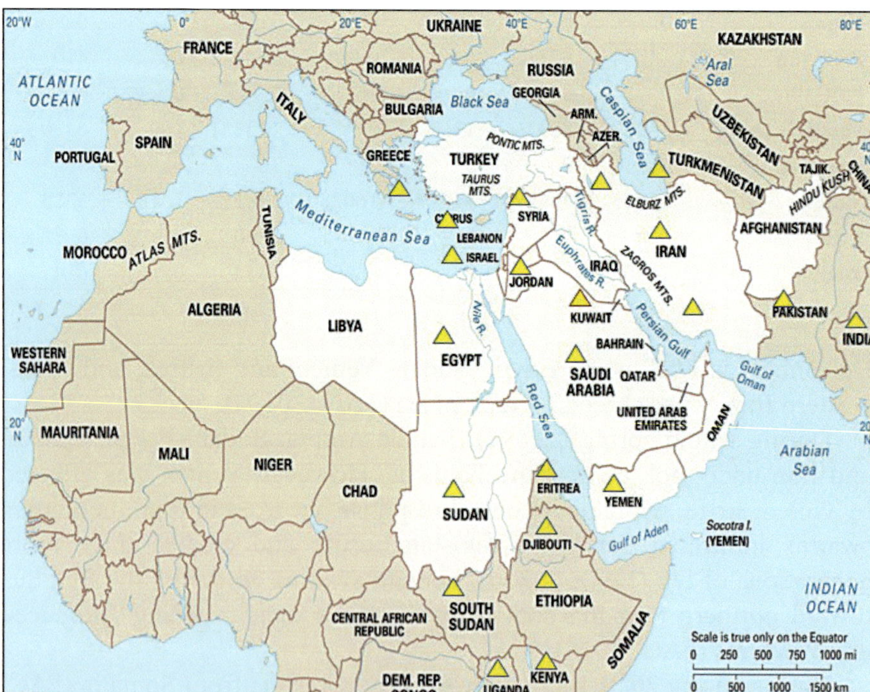

FIGURE 2.1 Desert locust infested areas during 2018–2021.

2.6 NEW APPROACHES

The issue of studying desert locust management without ethical concerns is illustrated by the Food and Agriculture Organization's (FAO) recommendation for an enhanced financial mechanism (Deshormes, 2011). The catalog provides a financial strategy aimed at several times of DL populace changing aspects; however, it does not contain any information regarding the source, that is, who should be accountable for giving/maintaining support or at what criteria obligation could be distributed equitably. No doubt, it is in complete descriptive form highlighting various techniques that are being done but contains little or no normative analysis, that is, what needs to be done in the case of locust attacks.

Similarly, the 2020 exchange between Tom, US Ambassador and Permanent Representative, United Nations Rome Based Agencies, and Metelerkamp, Research fellow at the Environmental Learning and Research Center Rhodes University, exemplifies intense disagreements over the goals of desert locust management. Despite opposing socioeconomic ideologies were offered, neither side in this argument went beyond basic normative declarations (i.e., whether industrial or agro-ecological values should be chosen) (Tom, 2021). As a result, political rhetoric mainly concealed the conflicting and compelling ethical grounds.

The present locust epidemic is a result of environmental and political factors (Lecoq, 2020), including military conflicts between weak nations in the Middle East and Eastern Africa, which prevented effective preventative efforts following high rains (Gay et al., 2018). Gay et al. (2019) discovered that a preventative program may be negated with merely 5% of the region having limited access using a spatially explicit multiagent model. According to the findings, plagues are more likely to start in weak states with broad territories that are unreachable to pest management.

When donor countries are hesitant to contribute assistance to countries that participate in terrorism and human-rights violations (e.g., Sudan), the risk of locust epidemics going unnoticed and unabated is high (Showler, 2019).

In addition, habitat mapping studies for the migratory locust have been done in various study locations in China using remote sensing. Based on a single Landsat TM (Thematic Mapper) picture, Liu et al. (2006) used a land cover classification-based methodology to generate the data of probable habitats of locusts in the Yellow River delta. Li et al. (2011) created a field cover categorization map and converted it to prospective habitats of Asian migratory locusts in Hebei Province using 14 HJ-1 CCD pictures (30 m spatial resolution) and NDVI time series.

In 2015, Zheng et al. (2018) used decision tree-based categorization to extract Oriental migratory locust habitat from six Landsat Operational Land Imager (OLI) photos in the Dongying area. MODIS and Landsat data were used to estimate yearly changes in oriental migratory locust habitat between 2000 and 2016. Zhao et al. (2020) have detected variations in land cover and land use in oriental migratory locust habitats across China.

Desert locusts are known to attack many crops, for instance, in Gujarat and Rajasthan, multiple attacks of locusts were reported in 2019 which destroyed wheat, potatoes, cumin, and rapeseed on more than area of 25000 hectares. Scientists from the Locust Warning Organization (Ganganagar, Jaisalmer districts) observed grasshoppers identified as desert locusts in 2020. The desert locusts seen were the same as migratory pests from East Africa (destroying wheat, maize, and sorghum crops) (Joshi et al., 2020).

2.7 CONTROL

Under cultural control, select the climatic resilient crop varieties and maintain the field that it should be free from weeds or crop residues by burning. Deep summer plough is to be done and recommended irrigation schedule is to be followed.

- Under chemical control, dusting with Methyl Parathion 2% @ 5–10% or 20–25 kg /ha./ Malathion 5% @ 20 kg /ha.
- Under biocontrol, entomopathogenic fungi (*Metarhizium acridum*) (or) bioinhibiting of pheromones (Guaiacol—especially for synthesis to attracting locusts and forming swarm) has been used through the application of *Pantoea agglomerans* (Zhang et al., 2019). Few other control methods are listed in Table 2.2.

2.8 MECHANISM OF OUTBREAK

During the recession phase, if unexpected rainfall occurs, the DL take the benefit and reproduce exponentially. Under normal circumstances, locusts will survive 3–6 months, and can multiply into next generations for about 16–20 times in their numbers. As soon as the desert environment tends to dry out, a huge number of DL abandon solitary lifestyle and collectively acquire physical contact, foraging in a unified group. As a result, it causes a chain reaction of changes in behavior and metabolism, resulting in trivial crowds of wingless nymphs and adults merging to create compact groups of hoppers along with adult swarms. During this process, their color changes from green to yellowish brown.

Gregarization is the term for this stage, and transient is the term for the period between solidarity and gregariousness, when locusts begin to form groups. Gregarization occurs only in recession area and breeding will result in two generations in a short period.

When locusts' number increases dramatically on a local scale as a result of their multiplication, concentration and gregarization, if not controlled timely, can result in hopper bands and adult swarms, subsequently leading to an outbreak (Roffey and Popov, 1968). Under circumstances of continued rainfall continuity, a significant amplification in DL population causes simultaneous outbreaks, which are accompanied by two or three seasons of transient to gregarious breeding in complementary seasonal breeding areas. This time is referred to as an upsurge.

The time of prevalence of DL (bands and swarms) occupying an expanding region for one or more years is usually referred to as plague. The occupying area is the product of numerous upsurges and plagues which have caused locust swarms to spread beyond recession zones, covering roughly 20% of the Earth's land surface (Cressman, 2016).

2.9 CLIMATE CHANGE IMPACT

Climate change is bringing a dynamic change affecting DL habitat, multiplication, relocation, and epidemic occurrence. It is well known that the current climate change is causing temperature to increase. Warmer temperatures will extend the length of the summer, winter, and spring reproduction time, also allowing DL eggs and wingless nymphs to grow earlier as long as this is associated with a continuation or increase of good rains. This is likely to be the most pronounced during the winter and may allow an extra generation of breeding to take place (Cressman, 2013).

Under warm temperatures, desert locust migration will be affected in such a manner that solitarious adults can fly up to longer distances in night, especially during the colder periods of the year. This can result in destination alteration whereby adults can reach their destination earlier or else swarms can reach new places that have not been reachable up to now. Climate change could also allow swarms to fly higher than 1800 m,

TABLE 2.2 Control Mechanism Adopted for Locus.

Inputs	Control Type	Remarks	References
Organochlorines	Chemical	Highly hazardous	
Argiope arcuata Lucas	Biological control	Predator for locust adults and nymphs	Pandey et al. (2021)
Bacillus thuringiensis Berliner	Biological control	Pathogen for adults	Pandey et al. (2021)
Blaesoxipha agrestis Robineau-Desvoidy	Biological control	Parasite for adults and nymphs	Pandey et al. (2021)
Gregarina garnhami Canning		Pathogen for adults and nymphs	Pandey et al. (2021)
Malamoeba locustae (King and Taylor)	Biological control	Pathogen for adults and nymphs	Pandey et al. (2021)
Nosema locustae Canning	Biological control	Pathogen for adults and nymphs	Pandey et al. (2021)
Pseudomonas aeruginosa Migula	Biological control	Pathogen for adults and nymphs	Pandey et al. (2021)
Sphex nivosus (Smith)	Biological control	Predator for adults and nymphs	Pandey et al. (2021)
Stomorhina lunata Fabricius	Biological control	Parasite for eggs	Pandey et al. (2021)
Systoechus aurifacies Greathead	Biological control	Predator for eggs	Pandey et al. (2021)
Metarhizium anisopliae var. acridum Driver & Milner	Biocontrol agent	Enhance acidic phosphatase results in autophagy	Xia et al. (2000)
Bacillus cereus (Bacteria)	Biocontrol agent	High insecticidal activity	Reda et al. (2018)
Metarhizium anisopliae	Entomopathogenic fungus	Reduced egg hatching	Bahia (2020)
Lambda cyhalothrin 2.5EC	Chemical control	79% mortality after 24 h @3 mL/ L 5-day-old winged adults)	Ahmad et al. (2020).
Deltamethrin 2.5EC	Chemical control	73% mortality after 24 h @3 mL/ L 5-day-old winged adults)	Ahmad et al. (2020).
Chlorpyrifos 40EC	Chemical control	77% mortality after 24 h @10 mL/ L 5-day-old winged adults)	Ahmad et al. (2020).
actinomycetes (Spinosad, Tracer 24%SC)	Bioagent	Successful for locusts and grasshoppers	Rada and Atreby (2017).
M. anisopliae var. acridum (Green Muscle)	Fungus	Controlled locusts and grasshoppers	Rada and Atreby (2017).

which is the general limit of flight due to temperature. If this is the case, then the Atlas Mountains in Northwestern Africa, the Hoggar Mountains of Algeria, the Jebel Akhdar Mountains in northern Oman, the mountains in the interior of Iran, and the mountain ranges along both sides of the Red Sea may no longer be natural barriers that impede migration. On the other hand, if warmer temperature regimes were to become extremely hot, for example, above 50°C, then desert locust presence and survival could become limited in some areas of the Sahara and the Arabian Peninsula (Meynard et al. 2017).

2.10 CONCLUSIONS

Desert locust, being one of the insects causing tremendous losses to crops and can take the form of plague, has emerged as a threat to food security, especially in recent years. The most affected countries fall in Arabian Peninsula, Southwest Asia, and East Africa. Moreover, the affected countries are already facing issues regarding food security and crop losses due to other factors. Hence, the locusts' attack can exacerbate the situation. Due to climate change, it is possible that locusts may attack other countries owing to seasonal changes. In this regard, well-planned management strategies are required with special emphasis on the early prediction of the attack, so farmers can prepare crops for control measures. Other alternatives like insect pest-resistant varieties need to be developed in future.

KEYWORDS

- **desert locust**
- ***Schistocerca gregaria***
- **plague**
- **natural disaster**
- **insecticides**
- **locust control**

REFERENCES

Ahmad, K. J.; Aslam, A.; Munir, M.; Ali, Q.; Hussain, D.; Malik, H.; Zubair, M. Toxicological Impact of Different Insecticides on the Desert Locust (Schistocerca gregaria Forsk.)(Acrididae). *Life Sci. J.* **2020,** *17* (8), 6–10.

Bahia, D. M. The Effect of the Entomopathogenic Fungus Metarhizium anisopliae on the Eggs of the Desert Locust Schistocerca gregaria (Forskål, 1775). **2020.**

Belayneh, Y. T. Acridid Pest Management in the Developing World: A Challenge to the Rural Population, a Dilemma to the International Community. *J. Orthoptera Res.* **2005,** *14* (2), 187–195.

Cressman, K. Climate Change and Locusts in the WANA Region. In *Climate Change and Food Security in West Asia and North Africa*; Springer: Dordrecht, 2013; pp 131–143.

Cressman, K. Desert Locust. *Biol. Environ. Hazards Risks Disasters* **2016,** 87–105.

Deshormes, A. Institutional Study to Enhance the Roles and Responsibilities of the Desert Locust Control Commissions Established Under Article XIV. *Financial Governance Final Report*, 2011.

Doré, A.; Ould Babah, M. A.; Barbier, M.; Lecoq, M. Rethinking Organization and Governance for Surveillance and Risk Management of Desert Locust Outbreaks, 2005.

FAO (Food and Agriculture Organization of the United Nations). Global Information and Early Warning System on Food and Agriculture World Food Program: Special Report, FAO/WFP Crop and Food Supply Assessment Mission to Morocco with Special Focus on Losses Due to the Desert Locust**,** 21 Dec 2004.

FAO (Food and Agriculture Organization of the United Nations). *Towards a more effective response to desert locusts and their impacts on food security, livelihoods and poverty***;** FAO: Rome, Italy, 2006.

FAO (Food and Agriculture Organization of the United Nations). Evaluation Final de la Phase II (2014–2017) du Programme EMPRES/Composante Criquet Pèlerin en Région Occidentale (EMPRES-RO); FAO: Rome, Italy, 2018. http://www.fao.org/3/ca4005fr/ca4005fr.pdf (accessed on 5 Nov 2020).

FAO (Food and Agriculture Organization of the United Nations). Desert Locust Bulletin, 9 Jan 2019; No. 483; FAO: Rome, Italy, 2019a. http://www.fao.org/ag/locusts/common/ecg/2455/en/DL483e.pdf (accessed on 4 June 2021).

FAO (Food and Agriculture Organization of the United Nations). Desert Locust Bulletin, 3 April 2019; No. 486; FAO: Rome, Italy, 2019b. http://www.fao.org/ag/locusts/common/ecg/2471/en/DL486e.pdf (accessed on 3 June 2021).

FAO (Food and Agriculture Organization of the United Nations). Desert Locust Bulletin, 3 May 2019; No. 487; FAO: Rome, Italy, 2019c. http://www.fao.org/ag/locusts/common/ecg/2472/en/DL487e.pdf (accessed on 3 June 2021).

FAO (Food and Agriculture Organization of the United Nations). Desert Locust Bulletin, 5 June 2019; No. 488; FAO: Rome, Italy, 2019d. http://www.fao.org/ag/locusts/common/ecg/2475/en/DL488e.pdf (accessed on 9 June 2021).

FAO (Food and Agriculture Organization of the United Nations). Desert Locust Bulletin, 3 Feb 2020; No. 496; FAO: Rome, Italy, 2020a. http://www.fao.org/ag/locusts/common/ecg/2526/en/DL496e.pdf (accessed on 1 June2021).

FAO (Food and Agriculture Organization of the United Nations). Desert Locust Bulletin, 5 Mar 2020; No. 497; FAO: Rome, Italy, 2020b. http://www.fao.org/ag/locusts/common/ecg/2531/en/DL497e.pdf (accessed on 1 June 2021).
FAO (Food and Agriculture Organization of the United Nations). Desert Locust Bulletin, 4 May 2020; No. 499; FAO: Rome, Italy, 2020c. http://www.fao.org/ag/locusts/common/ecg/2551/en/DL499e.pdf (accessed on 11 June 2021).
FAO (Food and Agriculture Organization of the United Nations). Current Upsurge (2019–2020). Desert Locust Watch, 2020d. http://www.fao.org/ag/locusts/en/info/2094/index.html (accessed on 3 June 2021).
FAO. *Impact of COVID-19 on Agriculture, Food Systems and Rural Livelihoods in Eastern Africa: Policy and Programmatic Options,* 2020e. https://reliefweb.int/report/burundi/impact-covid–19
FEWS (Famine Early Warning Systems). *FEWS Food Security and Weather Reports*; FAO: Rome, Italy, Dec 2004.
GAM. *East Africa 2020 Flood Impacts on Agriculture*, Updated 19 May 2020.
Gay, P. E.; Lecoq, M.; Piou, C. Improving Preventive Locust Management: Insights from a Multi-Agent Model. *Pest Manag. Sci.* **2018,** *74*, 46–58.
Gay, P.-E.; Lecoq, M.; Piou, C. The Limitations of Locust Preventive Management Faced with Spatial Uncertainty: Exploration with a Multi-Agent Model. *Pest Manag. Sci.* **2019,** *76*, 1094–1102.
IFRC. *East Africa: Red Cross Raises the Alarm Over a "Triple Menace" of Floods, COVID-19 and Locusts,* 2020. https://media.ifrc.org/ifrc/press-release/east-africa-red-cross-raises-alarm-triple-menace-floods-covid-19-locusts/
Joshi, M. J.; Raj, V. P.; Solanki, C. B.; Vaishali, V. B. Desert Locust (Schistocera gregaria F.) Outbreak in Gujarat (India). *Agric.Food: E-Newslett* **2020,** *2* (6), 691–693.
Kassegn, A.; Endris, E. Review on Socio-Economic Impacts of 'Triple Threats' of COVID-19, Desert Locusts, and Floods in East Africa: Evidence from Ethiopia. *Cogent Soc. Sci.* **2021,** *7* (1), 1885122.
Lecoq, M. Desert Locust Threat to Agricultural Development and Food Security and FAO/International Role in Its Control. Arab Society for Plant Protection, 2003.
Lecoq, M. Some Considerations of the Current Desert Locust Situation in East Africa. *Metaleptea* **2020,** *40*, 25–26.
Li, J.; Chen, J.; Sheng, S. Locust Habitats Monitoring Based on Multi-Temporal CCD Data of HJ-1 Satellite. In *MIPPR 2011: Multispectral Image Acquisition, Processing, and Analysis*. *Int. Soc. Optics Photon.* Dec **2011,** *8002*, 80021H.
Liu, Q.; Liu, G.; Yang, Y.; Liu, P.; Huang, J. Identifying the Breeding Areas of Locusts in the Yellow River Estuary Using Landsat ETM+ Imagery. In *Remote Sensing of the Environment: 15th National Symposium on Remote Sensing of China*, Vol. 6200; International Society for Optics and Photonics, June 2006; p 62000G.
Maeno, K. O.; Ould Babah Ebbe, M. A. Aggregation Site Choice by Gregarious Nymphs of the Desert Locust, *Schistocerca gregaria*, in the Sahara Desert of Mauritania. *Insects.* **2018,** *9*, 99.
Maeno, K. O.; Ould Ely, S.; Ould Mohamed, S.; Jaavar, M. E. H.; Nakamura, S.; Ould Babah Ebbe, M. A. Behavioral Plasticity in Anti-Predator Defense in the Desert Locust. *J. Arid. Environ.* **2018,** *158*, 47–50.

Maeno, K. O.; Ould Ely, S.; Ould Mohamed, S.; Jaavar, M. E. H.; Nakamura, S.; Ould Babah Ebbe, M. A. Defence Tactics Cycle with Diel Microhabitat Choice and Body Temperature in the Desert Locust, *Schistocerca gregaria*. *Ethology* **2019,** *125*, 250–261.

Maeno, K. O.; Ould Ely, S.; Ould Mohamed, S.; Jaavar, M. E. H.; Ould Babah Ebbe, M. A. Adult Desert Locust Swarms, Schistocerca gregaria, Preferentially Roost in the Tallest Plants at Any Given Site in the Sahara Desert. *Agronomy* **2020,** *10*, 1923.

Meynard, C. N.; Gay, P. E.; Lecoq, M.; Foucart, A.; Piou, C.; Chapuis, M. P. Climate-Driven Geographic Distribution of the Desert Locust During Recession Periods: Subspecies' Niche Differentiation and Relative Risks Under Scenarios of Climate Change. *Glob. Change Biol.* **2017,** *23*, 4739–4749.

Pandey, M.; Suwal, B.; Kayastha, P.; Suwal, G.; Khanal, D. Desert Locust Invasion in Nepal and Possible Management Strategies: A Review. *J. Agric. Food Res.* **2021,** 100166.

Pener, M. P.; Simpson, S. J. Locust Phase Polyphenism: An Update. *Adv Insect Phys*. **2009,** *36*, 1–272.

Rada,W. M.; Atreby, E. L. Efficiency of Three Natural Products for Controling Desert Locust, *Schistocerca gregaria* (forsk.) (Orthoptera: acrididae). *J. Environ. Sci. Int.* **2017,** *38* (2), 93–104.

Rahaman, M. M. D.; Saha, O.; Rakhi, N. N.; Chowdhury, M. G.; Sammonds, P.; Kamal, A. M. Overlapping of Locust Swarms with COVID-19 Pandemic: A Cascading Disaster for Africa. *Pathog. Glob. Health.* **2020,** *114* (8), 1–2.

Reda, M.; Mashtoly, T. A.; El-Zemaity, M. S.; Abolmaaty, A.; Abdelatef, G. M.; Marzouk, A. A.; Susceptibility of Desert Locust, *Schistocerca gregaria* (Orthoptera: Acrididae) to Bacillus Cereus Isolated from Egypt. *Arab Univ. J. Agric. Sci*. **2018,** *26*, 725–734.

Roffey, J.; Popov, G. Environmental and Behavioural Processes in a Desert Locust Outbreak. *Nature* **1968,** *219* (5153), 446–450.

Showler, A. T. The Desert Locust in Africa and Western Asia: Complexities of War, Politics, Perilous Terrain, and Development; Radcliffe's IPM World Textbook, 2009.

Showler, A. T. Desert Locust Control: The Effectiveness of Proactive Interventions and the Goal of Outbreak Prevention. *Am. Entomol.* **2019,** *65* (3), 180–191.

Steedman, A. *Locust Handbook: Overseas Development*, 2nd ed; Natural Resource Institute London, 1988.

Tom, K. The UN Should Learn That Ideology Won't Stop a Plague of Locusts. Real Clear World Website. https://www. realclearworld. com/2020/08/06/un_ideology_wont_stop_a_plague_of_locusts_501135. html (accessed on 11 January 2021).

Xia, Y.; Dean, P.; Judge, A. J.; Gillespie, J. P.; Clarkson, J. M.; Charnley, A. K. Acid Phosphatases in the Haemolymph of the Desert Locust, *Schistocerca gregaria*, Infected with the Entomopathogenic Fungus Metarhizium Anisopliae. *J. Insect Physiol.* **2000,** *46*, 1249–1257.

Zhao, L.; Huang, W.; Chen, J.; Dong, Y.; Ren, B.; Geng, Y. Land Use/Cover Changes in the Oriental Migratory Locust Area of China: Implications for Ecological Control and Monitoring of Locust Area. Agric. *Ecosyst. Environ.* **2020,** *303*, 107110.

Zheng, X.; Huang, J.; Li, H.; Mansaray, L. R.; Song, P.; Dou, Y. Mapping of Oriental Migratory Locust Habitat Using Landsat OLI Images in Dongying City, China. In *2018 7th International Conference on Agro-geoinformatics (Agro-geoinformatics)*; IEEE, Aug 2018; pp 1–5.

CHAPTER 3

Traditional Control and Management Technologies Used by Various Countries for Migratory Locusts

LAILA SHAHZAD[1,3*], AYESHA AMIR[1], ASMA YASIN[1], and WAJIHA ANUM[2]

[1]*Sustainable Development Study Center, Government College University, Lahore, Pakistan*

[2]*Regional Agricultural research Institute, Bahawalpur, Pakistan*

[3]*University College London, United Kingdom.*

ABSTRACT

The locusts' invasion is a growing aspect of traditional and scientific aspects of life which is much devastating for the crops and is a major threat to the agriculture of nations. It is one of the ancient threats, a vast area of land was damaged by the locusts. It affects the vegetation, pastures, forests, wild plants, and also the cultivated plants. The destruction of crops results in a series of chain reactions which result in famine, diversion of labor, disruption of trade, and cultivation. It also causes heavy expenditure on control measures. This chapter has tried to cover traditional as well as modern methods which are followed in different African and Asian countries. It also covers the new technologies to be intervened for the control of locust and its management, such as geographical information system and global positioning system. Robust policy interpolation is needed to control this havoc in the coming times.

Locust Outbreaks: Management and the World Economy. Umair Riaz, Khalid Rehman Hakeem, (Eds.)

3.1 LOCUST AS A CHARACTERISTIC SPECIES

Locusts belong to the group of grasshoppers; they are short-horned species that have a swarming phase. They are mostly solitary, but sometimes, they become gregarious due to change in habits and behavior. Though locusts are low in number, they have a huge impact on agriculture and have a major threat to the economy. But, in drought conditions and in rapid vegetation, they started to grow and become gregarious. These locusts are migratory and can travel long distances; they also cause damage to crops.

These locusts are widely spread on all continents except Antarctica and North America. They are mostly observed in North Africa, Australia, Middle East, and Indian Subcontinent. The locusts travel great distances. The outbreak of locusts was observed in Mali, Niger, Mauritania, and Sudan in 2003. Then they travel toward Morocco, Algeria, Africa, Egypt, Jordan, and Israel. Locusts had caused damage to about 50 countries in the world in one travel outbreak. The migratory locusts are classified into different sub-species mostly observed in Africa, Asia, Australia, and New Zealand (REF). The invasion of locusts is dramatic as it covers a large area in a very short amount of time. The invasion almost destroys all the green paths. The locusts cause destruction on a massive scale because they are in large groups and also cross the international boundaries. That is why the invasion of locusts attracts public and international attention (Tu et al., 2015).

The seasonal migration is greatly influenced by the climatic factors. The factors include wind, temperature, and vegetation. The locusts use downwind air for their flight. Locust outbreaks are the result of the area which becomes favorable for them. They breed in lush green vegetation and rainy weather condition. According to the metrological reasons, the duration of the locust outbreak may vary. It may occur at regular intervals. The locusts are invertebrate and have a very high migratory habit. They have voracious feeding behavior and marked polymorphism. They have the nature to take advantage of the geography and climate. They can survive in the temperature range between 0 and 60 degrees. They also have the ability to slow down or speed up their life cycle. Locusts can breed when they have favorable conditions. They can distribute in any part of the area. They are one of the very difficult insects to control due to diversity in the distribution area. They can utilize the environmental conditions and migratory nature for their distribution. Their migratory distance could be about thousand kilometers (Wei et al., 2017).

3.2 INTRODUCTION TO LOCUST CONTROL

In the recent 15 years, locust control in the world and especially in Africa has become a considerable type of polemic. Different questions concerning the locust attack, the impacts of insecticides used, preventative control strategy, and social as well as the economic importance of the locusts. Different projects were initiated in order to improve the survey and find some solutions and control methods for the treatment of chemical pesticides. Some most recent surveys have also been revised. Application techniques and surveys showed improvements with the help of GIS, precision spraying, and GPS technology. Different barriers are also available, this is just because of some persistent pesticides. Biological control with the aid of mycopesticides is very promising and it is the method that will gradually be followed. The atmospheric features are basically taken into an improved version, and it is an *integrated pest management* (IPM) technique which is now possible for the control of locusts. The Emergency Prevention System (EMPRES) was established by the FAO for transboundary animal and plant pests and diseases. This program has two gears: the plant pest part which focuses on the desert locust and the animal diseases part focuses primarily on rinderpest including five epidemic diseases (FAO, 2015).

In many different places, the local research capabilities are in the process of continual improvement. However, several important aspects actually remain a mystery. Among all the others, the basic concern is the sustainability of preventative control systems. Different countries that are present in this list are among the deprived in the world. These poor countries do not have necessary financial resources in order to fund the intensive type of control campaigns; they cannot even ensure proper prevention (Geng et al., 2020). The most realistic type of solution in order to prevent the locust plagues actually requires very strong commitment by the regulatory bodies and different donors, especially the establishment of the most effective emergency type of action plan, formulation of different complementary types of resources in an upsurge, and ready to quickly mobilize. Different types of emergency funds are necessary and they must be created. Rather than scientific or technological innovations, locust control currently depends more on political as well as on institutional choices.

3.3 HISTORIC KNOWLEDGE ON CONTROL

If we consider the history of humans, they have always been affected by the locust plagues. In the past centuries of human beings, this proved to be a recurrent theme for different travelers, naturalists, and missionaries. They are all actually a big witness to the severity of this major problem and the locust plague ultimately affects the supplies of food in Africa. All the means for the sake of controlling this locust plague is actually a rudimentary: prayer, noise, and fire, etc. Even today, this is a common human practice. An organized practice for the plague of locusts started when all the agricultural resources were misused in different countries of the world. The foremost scientific type of study on the subject and movements was carried out in Algeria, which proved to be very helpful in order to enhance the international type of awareness (Wei et al., 2017). The establishment of this sort of international cooperation and all the resultant progresses are basically renowned). In the 19th century, the scientific knowledge about the locust control was enhanced, and apparently, modern techniques appeared. These methods were then continually improved. The utmost development was attained between 1935 and 1950. Decades ago, a very good understanding of problems related to locusts in Africa and continual improvement concerning its management was actually achieved in both views technical as well as institutional. During this period, they experienced setting up of regional as well as national locust organizations, the establishment of international type of cooperations as well as the implementation of a unique preventative strategy. The current situation is a kind of paradoxical. We neither had too much scientific knowledge on different causes of the locust plagues nor such sophisticated ways of control of locust plagues. In the last years, we have seen that locust control in the region of Africa is frequently questioned. Some people in donor countries actually estimate that many key problems in locust control are basically due to not having sufficient documentation. In the 1980s, a significant locust plague took African countries with grasshopper outbreaks which were followed by the plague of the desert locusts. More lately, an outbreak of migratory Red locust badly affected Madagascar. As a consequence of those events, techniques for handling blighter locust populations, the extent of the locust downside, and also the effectiveness of locust management operations were questioned (Latchininsky et al., 2016).

The two broad methods of locust control include plague suppression and plague prevention. The first is an "effective strategy," which is used as an emergency, yet is a substitute measure in the event of a crisis and when locusts and hopper groups are being dispersed away from their original breeding grounds. This effective strategy is divided into "therapeutic" and "mitigation," in terms of eradicating the disease by simply controlling or simply trying to protect endangered distribution areas. Plague protection has three components: bump prevention, climb prevention, and upsurge eradication. With regard to the desert locust, the current strategy is a "preventive" approach. Specifically, it is a strategy to prevent climbing. It seeks to achieve the prevention of disease by reducing the population by trying to eradicate the emergence of the first human couple. These localized outbreaks occur during the economic downturn or should be the beginning of a disease sequence. Inaccessibility, lack of resources, and slow response, however, have led to performance and control of operations in all upsurges (Latchininsky et al., 2016). Therefore, the current "prevention strategy" is in fact a combination of outbreaks and monitoring of ups and downs, which rely on early control activities.

The strategy for desert locust considers effective international cooperation in terms of monitoring and control. Thus, there is a research-based prevention strategy that began about 100 years ago. This strategy has been used for about 40 years mainly by the Food and Agricultural Organization. Since the 1960s—a time when effective control measures were in place and prevention was in place—we could see a dramatic decline in the number of plagues. This fact is often cited as evidence of achievement in preventative management, although the impact of climate change and extreme drought in Africa at this time should not be overlooked (Latchininsky et al., 2016). Figure 3.1 is an effort to show the methods of locust control available today which were derived from the past knowledge.

3.4 AN OVERVIEW OF ASIAN COUNTRIES' STRATEGIES FOR LOCUST CONTROL AND MANAGEMENT

Asian countries, especially India, Pakistan, Nepal, and Iran have been widely studied for the locust attack and its management, but the major focus of research remains in African countries due to their greater effort. The farmers have nothing to do when they face grasshoppers or locusts.

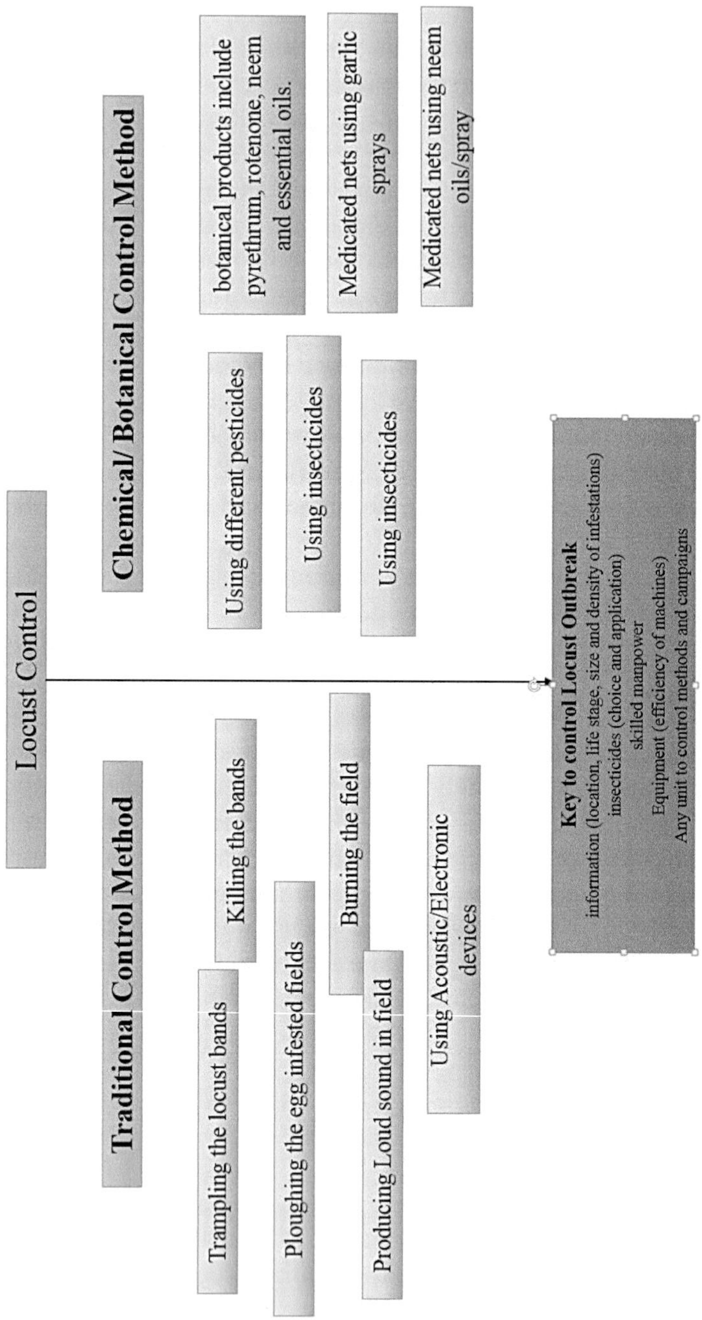

FIGURE 3.1 Methods and key concept of locust control.

They try a variety of physical and cultural control methods before following the chemical ones. Both the physical and cultural controls are used alone or in combination with the chemical ones. Most of the traditional methods are being replaced with the chemical methods. The chemical method was first used in India (Aryal et al., 2020). After this, chemical methods are modified with time, and now, aerial spraying has been used in which persistent organochlorines and several other insecticides are used. The BHC (benzene hexachloride) was then used as a traditional insect killer all over the world. Later, due to water shortage, the Exhaust Nozzle Sprayer was invented to control the locusts. Dieldrin was also used in 1960s for the control of locusts. DDT was also used but then it was banned in 1972 by the United States. The biopesticides were then developed from the fungus, which were used for the control of locusts effectively in Africa and Australia, but not used in India for the control of locusts. Some other methods used by the farmers for the control of the locusts are trampling and beating. Other methods include ploughing of fields or digging up, scattering the straws and also burning it, and lighting of the fires and making the noise, and use of the flamethrowers.

3.4.1 INDIA

India has faced several locusts' invasions and incursions in the past two decades. Locust swarms have the capacity to destroy an average as much food in one day as about 10 elephants, 25 camels or 2500 people can survive on (FAO, 2015). They not only destroy the harvesting crops but also cause damage to the flowers, fruits, seeds, barks as they devour the growing points of such food and non-food crops. They settle down in masses causing serious aftereffects.

Table 3.1 presents the information on locusts' attack in India.

3.5 MANAGEMENT AND CONTROL OF LOCUST

Several techniques were employed to manage locusts' swarms. The basic idea evolving since colonial times was to completely destroy their breeding grounds before they could again become a threat. A very popular technique that was followed in the country was the use of Cyprus screens which were oil-tarred screens meant for killing locusts, but it was proved

TABLE 3.1 Record of Last 50 Years of Locusts' Incursion in India.

Year	No. of swarms incursion
1964	004
1968	167
1970	002
1973	006
1974	006
1975	019
1976	002
1978	020
1983	026
1986	003
1989	015
1993	172
1997	004

Source: PPQS (2016).

to be not so effective later on. Other popular methods were the use of net system and dhotar method. The net system incorporates a capricious bag which was swung around the fields to trap young locusts. The dhotar method involves using a blanket to trip locusts resting on bushes. The main limitation of such orthodox techniques was that these methods demand huge manpower. Majority of Indians have a mind-set of these attacks as a "heaven-sent visitation" and they think that their god would take care of it (Das and Giri, 2020),

Now public participation has become a norm making it a job for people to collect them. It was kind of a relief job and they were paid accordingly. An insect-control technique involved the farmers ploughing the fallow land where locusts were resting; the escaping insects became the target for birds as birds could prey and kill more locusts than humans. In 1950s, ground and aerial spraying technique were adopted and were found highly effective till now. Insecticides used were organochlorines BHC and dieldrin (Sharma, 2014). Some other traditional methods which are used by the farmers for the control of the adult desert locust and hopper populations include (Uvarov, 1928) the use of flamethrowers, hoppers are beaten or trampled, excavation of egg pods or ploughing fields infested with egg pods, sprinkling straw over settling sites and then burning it.

Other methods include illumination fires or creation of noise to prevent swarms from settling in crops. Lastly, lashing hoppers into channels and burning or crushing them.

3.5.1 NEPAL

Nepal faced severe desert locust invasion in 1906, 1930, 1953, 1962, and recently in 2020 as recorded by the FAO (Pokhrel, 2020; Aryal et al., 2020). Possible management techniques were developed by the responsible government of Nepal to prevent any huge loss of crops and plants from the swarms of locusts when they attack. One of the common practices observed among the Nepali people is that they make loud noise using speakers or beating utensils against something hard. They believed it as a traditional way to cast away locusts' attack. This drill has no scientific grounds and mitigation of such huge attacks of locusts is not under the handling capacity of some farmers. Therefore, the government steps up and intervenes to present an effective remedy for desert locust invasion (Long et al., 2019).

3.6 MANAGEMENT AND CONTROL

Most widely used locusts control strategy is the application of ultra-low volume (ULV) technique in which pesticide is sprayed on the locusts in meager amount in the form of tiny droplets of concentrated pesticide formula (Van Huis et al., 2007). Greater use of pesticides on crop fields in far from sustainable strategy and the total efficacy of pest control ultimately decreases, therefore, serious concerns are being pointed out by the Integrated Pest Management to shift toward biological and other cultural techniques which are more sustainable (Matthew, 1999).

Biological control: use of various fungal pesticides as well as viral pathogens has been proved to be effective in locusts' control. But it bears a limitation that biopesticides takes longer time to act, so the results one need could be observed in long run. Thus, chemical pesticides are used for urgent use (Lomer et al., 2001).

Chemical control: In extreme emergency, when the locusts swarms are too much to handle, then chemical approaches involving harmful pesticides were the go-to approach in the past few decades. Organophosphate (DDT) chemicals are found to be most effective in controlling locusts' attack and terminating them. They are usually sprayed aerially, but hand-held sprayers can also be used (FAO, 2006). For this method to work effectively, it is important that locusts must come in contact with these pesticides. No instance of the development of any kind of resistance against such pesticides has been reported till now. For 1 hectare of land, only 0.5–5 L of pesticides are required (Aryal et al., 2020). Farmers on the other hand, prefer Dieldrin across the world (Lomer, 2001).

3.6.1 PAKISTAN

Since Pakistan lies on the migratory route of locusts' swarms, therefore, they are unfortunate to receive such locusts' attack both from the west side from Iran in Balocishtan. Similarly, locusts coming from India attack Cholistan Desert and Tharparkar Deserts. Pakistan faced first locusts' attack in 1926, while the worst among all was in 1952 with huge level of destruction, as the single swarm size was greater than 5 km long and 3 km wide (Boyang, 2021).

3.7 MANAGEMENT AND CONTROL

3.7.1 CULTURAL CONTROL

- Cultivation of soil where locusts' eggs were laid. Exposing them would ultimately dry it out or could be eaten by birds (Showler, 1995).

3.7.2 MECHANICAL CONTROL

- Insect suckers or other vacuum-like machines can be used to collect locusts as it will suck them from the ground.
- Use of flamethrowers to kill them altogether.
- Use of rollers to crush them.

3.7.3 BIOLOGICAL CONTROL

- Different types of beetles (blister beetle, ground beetle) and crickets are eggs predators.
- Using ducks to control locust.

3.7.4 MICROBIAL CONTROL

- Naturally occurring fungus *Metarhizium anisopliae.*
- *Metarhizium acridum* for killing both hoppers and adult locusts.

3.7.5 CHEMICAL CONTROL

Primary method of controlling locust swarms and hopper bands is mainly with broad-spectrum organophosphate agrochemicals (stomach and contact) applied in small concentrated doses, referred to as ULV formulations by vehicle-mounted and aerial sprayers and to a lesser extent by knapsack and hand-held sprayers. Application methods include sprays, baiting, and dusting. Other chemicals include: Chlorpyrifos 40EC (20 mL/1 L water), Deltamethrin 2.5EC (3 mL/1 L water), Cypermethrin (5% solution), Fenpropathrin (5% solution), Lambda-cyhalothrin 2.5EC (3 mL/1 L water), Bifenthrin (5% solution), and Carbaryl 85WP (dusting of mixture of 25 kg sand and 1 kg carbaryl) (Long et al., 2019).

The Table 3.2 depicts the land treated after locusts' attack in Asian countries.

TABLE 3.2 Land Treated During the April and May Locust Upsurge 2020.

Countries	Land treated in April (ha)	Land treated in May (ha)
Afghanistan	20	Not available data
Pakistan	50,289	76,466
India	1970	53,604
Iran	98,658	101,138
Iraq	815	101

Source: FAO (2020a, 2020b).

3.8 OVERVIEW AND MANAGEMENT IN AFRICAN COUNTRIES

Africa has a dry environmental condition which favors different pest attacks on growing crops. Developments in Africa and in Madagascar support the control as well as modern methods, to the detriment of change in the climate theory. The early 1970s lead toward better knowledge on the particular subject and preventative type of the strategy which is considered a more effective system for monitoring and an updated control method. Additionally, the use of ULV and organochlorine type of insecticide has proven to be a more effective approach. Thus, improvements were observed in few years' timeframe, yet operational problems arose. In Africa, in the last 16 years, locusts' attacks emerged again from 1988 to 1990, a Desert Locust plague affected several countries (Showler, 1995).

Locusts in the Red Sea region migrated to Sahel and North West Africa, following year-long winds and rains. In September 1998, large numbers of people crossed Atlantic to Caribbean and then in the northern part of the South America. Between 1998 and 2001, this was the case in Madagascar's Migratory Locust within 5 years, the country was attacked; even the western forest area was actually affected, and only the far North of the island survived from the attack of large locusts. In 1999, the average hemorrhage was 4001 ha. Extensive control operations are carried out tirelessly for 4 years with strong wind channels and vital international assistance, and apparently, extensive use of pesticides; this has had the consequences for the Malagasy region in living species and highly preserved (Showler, 1995).

Globally, the cost of fighting these plagues was enormous. More than "US $ 350 million" for Desert Locust, and more than US $ 60 million for "*Malagasy Migratory Locust.*". Large areas of land needed to be managed (incomparable to what could be managed by prevention). Between 1987 and 1999, 28 million ha were managed against the Western Hemisphere in 29 countries. "*Dieldrin*" inhibition required complete spraying, which is often used repeatedly and with insecticides that can be contraindicated such as "*fenitrothion*" and "*malathion.*" Such treatments may have a greater impact on the environment than the "*organochlorine*"-based treatments they have used. In Madagascar, more than 4.9 million hectares were treated during the recent epidemic (1.9 is fully covered with deltamethrin, propoxur, and fenitrothion and the remaining 8 million ha in "fipronil" prevention applications).

What are the actual reasons for the return of suffering and what lessons can we learn from them? Is it actually a failure of different strategies, or the inappropriate type of control measures or there is a lack of knowledge about the basic problem? The answer is straightforward: it is the consequence of a long period of forgiveness. In both Africa as well as in Madagascar, since the period of 1960s, a greater locust calming has actually led us to a decline in the support for regional management organizations as well as national locust control organizations. The latest plagues were actually foretold in advance. But the response capabilities of different countries have actually been declined making it difficult to actually stop rising once they start. After that they had to accept the difficulty of the situation, there is a need to do something about control due to the lack of alternatives, by going through problems like product selection and then integrating the necessary international resources. When all the problems were met, the problem had already been resolved and there was a plague. Africa met with a similar situation in 1988 as Madagascar. Clearly, there are some other reasons that we can put forward, especially with reference to the desert locust: security issues, terrorism, wars, etc., make access to certain places difficult or not possible to reach (Ma et al., 2012). A fundamental issue remained the global collapse of the governing bodies. The result of the recurrence of the disease, and especially this locust swarm first in 1988, was present in the global inquiries largely that was influenced by the so-called donor countries that funded during times of disease.

Among many people, questions are usually focused on the following:

Significance of locusts economically (many commitments were made based on Governmental and spiritual thoughts but not on the danger evaluation off food safety as well as agricultural production).

- The inhibitory plan (which is being criticized due to absence of efficiency along its transparency).
- The perspective of the FAO regarding issue along guidance.
- A selection of all types of insecticides that greatly influence the environment.
- The effectiveness as well as practicability of biological control strategies.
- Capability of different countries that play an important role in controlling their economic resource.
- Despite of all feasible substitute solutions, the selection of an accurate point of control.

Suitably or unsuitably, many people assume that in order to deal with the locust problem, the time had come to adopt a new strategy. The financial effect of acridids has become a matter in hand. Its objective was to evaluate its influence correctly and also to find out whether this influence is overstated or not, providing the striking nature of epidemic along the governmental pressure. It was also crucial to find out whether this defensive measure is feasible option or not.

In the present as well as in the past, we found many facts as well as verifications that demonstrate the financial significance of not only grasshoppers but also of locusts. It was also found that within only few hours, locusts have the ability of causing total reduction of the crops. But at present, these data are thought to be outmoded and therefore many questions rise on the financial significance of locusts. Although many people consider both grasshopper as well as locust as small insects but these small pests have ability to cause destructive damage to the localized crops in a short period. Furthermore, they also formed a model for the purpose to make comparison between the total cost of preventive measures and the loss prevented in the case of desert locusts. They also found that the action plan being implemented at present in terms of cost and benefit does not make any financial sense. Therefore, they ask for the re-evaluation of the current action plan to control locusts (Tu et al., 2015). Moreover, it was also observed that it is only the belief of the man that these small pests are of greater significance, but in reality, due to their small size, they only form a small part of the crop protection problem.

In addition, the recent low damage in Madagascar along Africa plays an important role in strengthening the above previous thoughts. For instance, during 1988, it was reported by the USA that desert locusts play an important role in causing loss of crops at small scale only, while on the other hand, it was also observed that the intensity of current epidemic is in fact high.

Although flaws are found in the role of an organization in controlling locusts' attack, as compared with the past, the current preventive measures are more effective. Not only the time duration of invasions of these insects is reduced but also, they are now controlled in a proper as well as a better way. Moreover, the locusts' attacks that occurred in the beginning of the 20th century cannot be compared with those that occurred in the 1960s, because at that time, the preventive measures are less effective and also because the intensity of the current attacks is less as compared with those that occurred in the past. However, at present, conflict between these two concepts is continuing. Furthermore, the countries that are currently affected by the attack of these pests consider this problem a greater issue. Proper evaluation of loss caused by locusts along with consideration of preventive measure is required by those countries that take part in the control of these insects (Roy, 1979).

Although some of the methodologies to deal with the problem are present, it is hard to properly find out the damage caused by these pests. In maintaining agricultural societies, it is necessary to consider both cash value and the collective value of the crops. In the case of these insects, the cost-benefit relationship fails, because the preventive measures are funded by different regions for the purpose to facilitate the remote areas. It is also crucial to evaluate both possible as well as the total loss that is prevented with the help of preventive measures. Due to the above hurdles, some people begin to believe that proper evaluation of preventive measures is impossible in Africa.

Nevertheless, a long current study on this topic has also observed the evaluation of financial risks relating to the desert locus in a better way. Although it is observed that in any country, the loss cannot affect the global supply of food, but at small scale, its effect is extremely large. Moreover, it is also found that the chance of yearly loss is very high and found to be in US dollars of tens of millions. While in the present models, the loss to the fields cannot be considered properly. However, in economic terms, the cost of lost harvesting is very low but the impact of these epidemics is found to be very severe in some regions and greatly affects the economy of these areas. Therefore, it is necessary to provide greater importance to all the studies that are related to the financial effects of these small insects. It is also found that although many donor countries based their funding on the financial effects of the preventive measures that play a vital role in controlling these small insects, it continues as a sensitive matter (Latchininsky et al., 2016).

In Algeria, in order to justify the preventive measure as against these desert small insects, scientists made contributions at the end of the 19th century. Different studies show that in the 20th century, about five major epidemics were found and about 50 years of these epidemics were observed in one century. The issue remains unsolved. All people accept the accurate evaluation of danger, but from time to time, the financial issue continues to rise. In the end, it was observed that it is necessary to control these small pests properly regardless of the cost–benefit relationship, moreover, it is also crucial to consider economic, social, as well as political aspects in controlling these insects.

3.8.1 CONTROL PLAN TO DEAL WITH LOCUSTS

For many years, action plan to control locusts has remained the center of dispute among different countries. It is also observed that many talks were held to find out which action plan is better in a way that it plays an important role in reducing both the environmental as well as financial value of controlling these small insects. And at present, on both grounds, theoretical as well as practical, this action plan was questioned by many people. Following are some questions that are raised on the liability of this action plan:

- Can a survey of huge deserted areas is possible in this situation?
- Is it possible for the contributor countries to continue assisting the preventive strategies?
- Can we have any information regarding the effectiveness of control measures?
- Will appropriate assets be available for the concerned countries to take possible preventive measures as well as to conduct a survey of the affected areas?
- Pre-planned positioning of resources?
- Detection of cluster of these small insects?

It is observed that the control measures suggested by the FAO are similar to these approaches and usually applied in Madagascar or in the Western areas that are habitat of these desert locusts. From many arguments on the topic of an action plan, it is proposed to reinforce the monitoring of these insects at the national level or to develop such control operations that play an important role in dealing with the population of these locusts properly and sustainably. From the past events, it can easily be observed that the influences of an epidemic can increase greatly if we do not take any preventive measures to control it. Moreover, by comparing current preventive strategies, we can find out that the preventive measures that are taken during an epidemic are in direct relation to the action plan which was adopted in the early breeding areas (Lecoq, 2001). At last, from many years of debates, it was observed that the most efficient option is still the control operations, because this strategy does not produce any negative impact on environment, and is cheap and also it plays a vital role in developing as well as maintaining different frameworks along skills in different countries. This issue is of greater significance because it not only helps the above countries to avoid their dependency on any external funding but it also helps these countries to tackle the issue of locusts.

3.8.2 TIMELY WARNING AND OBSERVATIONS

Currently, program related to advance warning has been found in the following disciplines:

- Communication
- GIS
- Space remote sensing
- Biogeography

The extended period plays an important role in obtaining a more accurate pictures of epidemic causing population, and this happened especially for the locust found in the deserted areas. Because, for such areas, examining is an essential part of the control operations. Improvements were made in the communication means to provide benefits to the preventive strategies to control locust's population. With the help of the internet, it became easy to transmit the situation of these insects in the form of data in any country. Moreover, the updates that are forecast by the Food and Agriculture Organization can easily be transferred in the form of e-mail. Also, the information on the internet regarding locusts is updated on regular basis. Moreover, it also became easy to integrate information regarding these small insects and their surroundings into the computer easily and properly, and this information can be used for efficient prediction of future problems to be caused by locusts. It is shown by different studies that the geographic information system was also developed to deal with the problems caused by migrant locusts found in Malagasy as well as desert locusts, and such a system is available in deserted areas in the form of two versions ***RAMSES and SWARMS***, where *Reconnaissance and management system for the environment of "schistocerca" (RAMSES)* system is available in simple version and can work easily for mobile computers that are found in all countries.

Schistocerca warning management system ***(SWARMS)***; On the other hand, this system plays an important role in covering the total reproducing areas of locusts found in deserted areas and this system is also properly dealt in the city of Rome by the Food and Agriculture Organization. "***GIS***" is the truly effective way that plays an important role in organizing control strategies as well as their examination (Lecoq, 2001).

The technology of remote sensing (RS) of space also plays an important role in this regard. For instance, satellites, such as NOAA and SPOT VEGETATION play an important role in observing those areas where better vegetation and rain are found, and such areas not only provide guidance to the teams that perform a survey of different grounds but also support the reproduction of these small insects.

Numerous issues obstruct the recognition of those reproducing areas of desert locusts where plant cover is found to be low. At the heart of small streams or canals, this type of plant cover is found to be very limited or scattered. Therefore, still many developments are required and the control strategies remain under many questions. Primarily, control strategies lie in the effective timely entry of data which are related to the environmental quality, and this happened in the case of both migrant locus of "***Malagasy***" as well as deserted locusts of deserted areas. Studies also show that accurate interpretation of the images is taken by satellite because the network on the ground is insufficient to bring such developments. It can only bring improvements. As we know that technology is progressing day by day and new types of sensors are emerging in the market, therefore irrespective of some current failures, it is necessary to continue the usage of the above technology in order to tackle the locusts' problem.

There is ban on organochlorines as well as on dieldrin, so locust control is basically based on the organophosphorus products, but now on the ban is on the pest regulators, such as dimilin and imidacloprid. The application techniques are more suitable for the implementation and particularly for the environment in "Madagascar" where there was recently an attack by locusts, and for the control measures, we could apply the new technology. The main purpose is to minimize the impact of the locust on the environment. The chemically installed way is creating concerns about the environmental protection. To control the locust, there is need to apply the proposed control measure at the patch area and then to implement it on a wider scale. Recently, a technique has been developed of "***GPS5***" in which with the help of the satellite movement, we can easily locate the location of the satellite. There are ways through which we can do this

- Locate desert-based locust easily.
- Well coordination with the help of which we can transfer aircraft.
- Desired area and precise attack.

The area of 1 hectare is used and targeted in the spread. With the recent passage of time, there are various hurdles in the way of implementation of the techniques. The use of chemicals, such as "*IGR*" and "*fipronil*" is a hurdle in the way of implementation. Therefore, preventive strategy needs to be implemented. Following are the advantages of using such type of applications we use:

- Prevention and control at a wider scale.
- Chemical use for per hectare area is lesser.
- The impact on nontargeted zones and areas is lesser.
- The impact on nontargeted fauna is also lesser.
- Economical in terms of time and insecticide spread.

So, there are recently announced methods that are beneficial in terms of long-lasting effects. Various natural ways could be used to curb the attack. The case studies on the chemical control methods are available in which it was observed that there was a reduction in the affected area as well as a reduction in the number of locusts and buds. There was a considerable decrease in the number of the locusts. This case study was of the Madagascar. Although there is wide range of benefits of the chemical control method, still it needs to go on the alternative methods.

3.8.3 ALTERNATIVE METHODS OF CONTROL IN OTHER PARTS OF THE WORLD

In botanical ways of control, vegetal extracts are used to kill the locust and control their growth. Although this is still in the initial stage, still it needs to be strengthened for use. This needs some improvement and needed to be applied on the larger patch of the affected area. "***Neurotransmitters***" act as semi-chemicals in the control of the locust (Tu et al., 2015). Pheromone could be used for this purpose at the transient stage. "***Pheromone***" control is a method to control pesticide and their production.

"***Mycopesticides***" are currently being used for the replacement of synthetic pesticides with the fungus-based pesticides. Recently, in Australia and southern part of America, there is oil production, which is a favorable way toward the control of locusts. These products can produce long-lasting influences, but are not suitable for immediate situations. They are presently being sold in the locust control market. Presently, only chemical and non-chemical-based pesticides are being used.

Authorities should not go for the rapid and alternative ways for pest control to enhance the agricultural production. The following factors may be considered:

- Proper providence of assistance.
- Antilocust policies for the farmer community.
- This strategy is not socially and economically viable for some of the countries.
- IPM (integrated pest management) approach.

Recently, it was adopted by some of the countries, and now the authorities are trying to apply this on desert and migratory locusts and control their population by integrated pest management. The most commonly used tools for this purpose include:

- Cross-sectional study of the population of an area.
- Monitoring and evaluation of the population.
- To understand the cost-effectiveness of the control method.
- Threshold.
- Study of diversity with relation to the environment.
- Use of biological method.
- Environmental impact of the environmental operations.
- Drones for desert locust detection and control.

To control the locust problem in Africa, integrated management system is being applied. The affected countries have their own locust control center (Latchininsky et al., 2016). Different commissions, companies, coordinations are developed for locust control management and monitoring their progress. Various institutional and political aspects are affecting the locust control commission. They are

- To control locust, some countries lack financial resources, and some lack trained officials.

- Countries like Africa are not fully funded to deal with locust population.
- Strategies for the crisis management.
- Long-term management response.

As far as locust population control is concerned, the most effective matter until now is that in 2000, a commission was made in which nine countries were involved for the locust control and in the formulation of strategy. This means that the things are moving in the right direction. At the same time, it is of great concern that things are not leading in the opposite direction and causing misinterest. The strategy of the locust control and possible misinterest in this context is the long-term consequences of the taxes and costs that countries have to bear for the control strategy. To ensure the sustainability, various methods are followed as given below:

- Commitments between the states.
- Locust control and less costly efforts.
- Locust situation and enforcement strategy.
- Emergency plans and strategy.
- Funds for the emergency situation.
- Cooperation at the technical and financial level.
- Proper emergency control system.

There is a major gap between the policy makers and donors, which can lead to new plague. Still on this point, the pest management strategy is unresolved (Geng et al., 2020). There are various issues in the implementation of coherent actions against locusts' attacks.

3.9 CONCLUSION

After many years of the locust calm situation, there is sharp locust invasion in many countries around the globe. Since the 1980s, there is straight rise in the locusts' attack which is demanding for a locust control policy. There is need for policymaking in this field. The economic impact of the locust attack is also being studied. There are various social concerns about the locust control and social constraints. Regarding the donors, the countries' social and economic condition matter a lot. However, it is needed for all the countries to implement the control strategies according to their demand and capacity. It is necessary for every country to document the situation in the given domains. The "***GPS***" involvement and implementation is a new way forward toward the betterment of the system. With the better navigation of the system, we can develop better and efficient ways for better networking and coverage of the system. Although there is large-scale development in the scientific and technical aspects, it still depends on the financial, institutional, political, social, and economic stability. There are also hurdles toward the implementation of the antilocust policy formulation. Investment in the relevant fields is necessary; otherwise, all efforts to control locust, particularly in developing countries would be in vain. It is necessary to develop political, social, and economic solution. Just the implementation of the technical solutions is not sufficient. Otherwise, the locust number would continue to surge, and it will cause the dire consequences for human and the environment, wrapping all social, economic, and political aspects.

KEYWORDS

- **locust control**
- **sustainability**
- **technological innovations**
- **traditional methods**

REFERENCES

Anil, S. Locust Control Management: Moving from Traditional to New Technologies—An Empirical Analysis. *Entomol Ornithol Herpetol.* **2014,** 4–1. DOI: 10.4172/2161–0983.1000141.

Arnold, V. H.; Keith, C.; Jouce, I. M. Preventing Desert Locust Plagues: Optimizing Management Interventions. *Entomologia Experimentalis et Applicata.* **2007,** *122*, 191–214. https://doi.org/10.1111/j.1570-7458.2006.00517.

Aryal, S.; Joshi, S. L.; Humagain, S.; Rajbhandari, R. D.; Acharya, M. C.; Subedi, R. K. Desert locust (Schistocerca gregaria) Identification, Characteristics and Management Journal Pre-proofmeasures, Harihar Bhawan, Lalitpur. 2020. http://www.npponepal.gov.np/content/211/2020/66540157/

Boyang, T. The Causes and Solutions Towards the Increasing Locusts Plague in Pakistan. *Web Conf.* **2021,** *228*, 02002. https://doi.org/10.1051/e3sconf/202122802002

FAO. Desert Locust Bulletin 499, 2020. https://reliefweb.int/sites/reliefweb.int/files/resources/DL499e.pdf.

FAO. Desert Locust Bulletin 500, 2020. https://reliefweb.int/sites/reliefweb.int/files/resources/DL502e.pdf

FAO. FAO Desert Locust Control Committee Thirty-Eighth Session, Rome, 2006. Journal Pre-proofhttp://www.fao.org/ag/locusts/common/ecg/1138_en_DLCC38e.pdf

FAO. FAO Desert Locust Information Service; FAO, 2015. http://www.fao.org/resilience/resources/resources-detail/en/c/278608/ (accessed September 7, 2020).

Geng, Y.; Longlong, Z.; Yingying, D.; Wenjiang, H.; Yue, S.; Yu, R.; Binyuan, R. Migratory Locust Habitat Analysis with PB-AHP Model Using Time-Series Satellite Images. *IEEE Acess* 2020, *8*, 166813–166823. DOI:10.1109/ACCESS.2020.3023264

https://www.preventionweb.net/news/view/48722.(n.d.)

Latchininsky, A.; Piou, C.; Franc, A.; Soti, V. Applications of Remote Sensing to Locust Management. *Land Surface Remote Sensing- Environment and Risks*, **2016;** pp 263–293.

Lecoq, M. Recent Progress in Desert Migratory Locust Management in Africa. Are Preventive Actions Possible? *J Orthoptera Res.* **2001,** *10* (3), 277–291. DOI:10.1665/1082-6467(2001)010[0277:RPIDAM]2.0.CO;2

Lomer, C. J.; Roy, B.; Daniel, L. J.; Juergen, L.; Matthew, B. T. Biological Control of Locusts and Grasshoppers. *Annu. Rev. Entomol.* **2001,** *46*, 667–702.

Long, Z.; Michel, L.; Alexandre, L.; David, H. Locust and Grasshopper Management, *Annu. Rev. Entomol.* 2019, *64*, 15–34. https://doi.org/10.1146/annurev-ento-011118- 112500.

Ma, C.; Yang, P.; Jiang, F.; Chapuis, M. P.; Shali, Y.; Sword, G. A.; Kang, L. Mitochondrial Genomes Reveal the Global Phytogeography and Dispersal Routes of the Migratory Locust. *Mol Ecol.* **2012,** *21* (17), 4344–58. doi: 10.1111/j.1365–294X.2012.05684.x. Epub 2012 Jun 28. PMID: 22738353.

Matthew, B. T. Ecological Approaches and the Development of "Truly Integrated" Pest Management. *Proc. Natl. Acad. Sci.* **1999,** *96* (11), 5944–5951. DOI:10.1073/pnas.96.11.5944

Pallavi, D.; Vineet, K. G. How Colonial India Fought Locusts Attack-and What We Could Learn from Those Tactics. 2020. https://scroll.in/article/963306/how-colonial-india-fought-locust-attacks-and-what-we-could-learn-from-those-tactics

Pokhrel, V. Locust: This Is How It Was Identified by Bringing It to the UK, a "New Grasshopper" Found in Nepal Nearly Six Decades Ago, BBC. 2020. https://www.bbc.com/nepali/news-52935683.

PPQS. Government of India Ministry of Agriculture and Farmers Welfare Department of Agriculture Cooperation and Farmers Welfare Directorate of Plant Protection, Quarantine Storage, 2016. http://ppqs.gov.in/divisions/locust-control-research

Roy, J. Decisive Steps Towards Control of the Desert Locust, 1952–62. *Phil. Trans. R. Soc. London. B Biol. Sci.* **1979,** *287* (1022), 301–304 (4 pages).

Showler, A. T. Locust (Orthoptera: Acrididae) Outbreak in Africa and Asia, 1992–1994: An Overview. *Am. Entomol.* **1995,** *41*, 179–185.

Showler, A. T.; Ould Babah Ebbe, M. A.; Lecoq, M.; Maeno, K. O. Early Intervention Against Desert Locusts: Current Proactive Approach and the Prospect of Sustainable Outbreak: Prevention. *Agronomy.* **2021,** *11*, 312. https://doi.org/10.3390/agronomy11020312

Symmons, P. M.; Cressman, K. Desert Locust Guidelines, Biology and Behavior. *Food Agric. Org. UN, Rome* **2001,** *2*, 2–25. http://www.fao.org/ag/LOCUSTS/common/ecg/347_en_DLG1e.pdf

Tu, X.; Wang, J.; Hao, K.; Whitman, D. W.; Fan, Y.; Cao, G.; Zhang, Z. Transcriptomic and Proteomic Analysis of Pre-diapause and Non-diapause Eggs of Migratory Locust, Locusta migratoria L. (Orthoptera: Acridoidea). *Sci. Rep.* **2015,** *5*, 11402. https://doi.org/10.1038/srep11402

Uvarov, B. P. *Locusts and Grasshoppers*; Imperial Bureau of Entomology: London, 1928; p 252.

Wei, J.; Shao, W.; Wang, X.; Ge, J.; Chen, X.; Yu, D.; Kang, L. Composition and Emission Dynamics of Migratory Locust Volatiles in Response to Changes in Developmental Stages and Population Density. *Insect Sci.* **2017,** *24* (1), 60–72. https://doi.org/10.1111/1744–7917.12396

CHAPTER 4

Emerging Strategies to Combat Locust Outbreaks

UZMA AZEEM[1], KHALID REHMAN HAKEEM[2], and M. ALI[3]

[1]*Sanmati Government College of Science Education and Research, Ludhiana, Punjab, India*

[2]*Department of Biological Sciences, King Abdul Aziz University, Jeddah, Kingdom of Saudi Arabia*

[3]*Department of Pharmacognosy,College of Pharmacy, Jazan University, Kingdom of Saudi Arabia*

ABSTRACT

Locusts (Orthoptera: Acrididae) are historically proven insect pests of agricultural crops growing around the globe. Their management is crucial to food security throughout the world that needs governmental/international participation. Locusts exhibit solitary phase at low population density and gregarious phase at high population density. Gregarious hoppers gather and march in dense bands, and adults swarm in large numbers to long distances causing huge damage to various major and minor crops, such as fruits, vegetables, legumes, and cereals across the globe. This damage to vegetation results in great economic losses. Several conventional methods have been employed for the management of locust outbreaks. However, these methods are expensive, less effective, and of short-term use. Furthermore, the use of insecticides poses threat to the natural enemies of locusts

Locust Outbreaks: Management and the World Economy. Umair Riaz, Khalid Rehman Hakeem, (Eds.)

as well as to the useful insects with eco-friendly roles. Therefore, novel, cost-effective, and eco-friendly strategies with long-term efficiency are urgently required. These strategies include remote sensing, use of latest monitoring technologies, biological control, biorational control, insect growth regulators, genetic control, integrated pest management, etc. Additionally, public participation, the establishment of regional organizations, national and international cooperation, legislation, and economic funds are also necessary to attain the desired outcomes.

4.1 INTRODUCTION

Locusts are short-horned grasshoppers classified in the Family Acrididae. A dozen out of the 6400 described species of grasshoppers show phase transition and are considered locusts. Locusts exhibit density-dependent changes in behavior, physiology, and also display phenotypic polymorphism. The solitarious and gregarious locusts show differences in their morphology, food habits, physiology (nutritional reproductive, neurophysiology, endocrinology, and pheromone production), life span, and even at molecular levels (Simpson and Sword, 2008; Pener and Simpson, 2009; Latchininsky, 2010). The ability to generate a swarming phase evolved independently many times in the Family Acrididae and is an advanced evolutionary feature (Song, 2011). In many holy books, such as the Torah, the Koran, and the Bible, locusts are mentioned several times as agricultural pests causing famines (Uvarov, 1944; Steedman, 1988; Kritsky, 1997). Locusts regularly display a phase transition from solitary phase to gregarious phase, reproduce profusely, migrate, inundate and damage crops, leading to a plague (Shi et al., 2014). In ancient Egypt, the locust plagues have been observed since Pharaonic times and are considered one of the most devastating plagues posing risks to agriculture and food security (Verlinden et al., 2020). *Locusta migratoria*, *Schistocerca gregaria*, *Locustana pardalina*, *S. piceifrons*, *Calliptamus italicus*, *Dociostaurus maroccanus*, *Nomadacris septemfasciata*, *S. cancellata* are the most destructive locust pests (Belayneh, 2005; Ramesh, 2015; Gall et al., 2019; Vijayalakshmia and Meenaa, 2020; Peng et al., 2020). History witnesses various plagues associated with locusts, such as the plague caused by *Chortoicetes terminifera* with widespread impact on agricultural areas in several states of Australia during 1933–1935, 1953–1955, 1973–1974,

1978–1979, 1983–1984, 1992–1994, 1999–2000, and 2004–2005 (Deveson, 2011), *L. migratoria migratoria* plague in Kazakhstan in 1999 damaged 2200 km^2 of grain crops costing USD 15 million (Latchininsky, 2013), *L. migratoria manilensis* (Chi, 1950; Yu et al., 2009), *L. migratoria capito* (Razafindranaivo, 1884), *D. maroccanus* plagues has caused crop damage in 25 countries, especially in Central Asia (Latchininsky, 1998), *C. italicus* plague had caused damage to agricultural crops in Russia, Afghanistan, and Pakistan (Latchininsky et al., 2002), *N. septemfasciata* plague is also known in Africa, Madagascar, Mauritius, and Reunion (COPR, 1982). The recent desert locust (*S. gregaria*) plague in large parts of Africa started in 2018 and has affected more than 30 nations (FAO, 2020). These divesting locust species feed on nearly 500 plant species, including grasses, agricultural crops, vegetables, fruits, and trees (Latchininsky, 2013; Ingrisch et al., 2009; Cressman and Locust, 2016). Swarms of locusts can occupy several hundred square kilometers with migration rate of up to 200 km/day. A swarm with approximately 40 million locusts per square kilometer can eat food per day equivalent to be taken in by about 35,000 people (Vreyer et al., 2012). These swarms produce significant long-term socioeconomic impacts. The damage to crop production is tragic for local farmers and results in hike in food prices in the local market that in turn affects families doing jobs other than farming. The poorest broods are generally the worst hit. Malnourishment of children and pregnant women poses danger to their growth and development. During locust plagues in Mali (1987–1989), school enrolment rate dropped to $1/4^{th}$, with girls enrolment being mainly affected (Courcoux, 2012). Anthropogenic activities, political relations among affected countries, water system change, and climate change affect the propensity of locusts to swarm (FAO, 2020; Calvão and Pessoa, 2015; Cullen et al., 2017; Meynard et al., 2020). Keeping in view the environmental, social, and economic losses brought about by locust plagues, there is an urgency to develop novel and more effective methods to keep check on locust outbreaks. Many conventional methods were in practice. However, these were more laborious, less effective, time-consuming, and of short-term effect. Recent scientific and technological developments prove beneficial to overcome all these shortcomings (Peng et al., 2020). Here in this chapter, recent advances in various strategies to fight locust outbreaks are discussed.

4.2 MONITORING AND FORECASTING

Currently, the monitoring and forecasting systems include several integrated components that are working evenly and assuredly to provide more precise and rapid information and to alert a wide international audience regularly (Zhang et al., 2019). Substantial improvements have been made in locust monitoring and early warning systems with the invention and utilization of new technologies in various streams, such as communication, computing, geospatial data, remote sensing, etc.

4.2.1 COMMUNICATION MODES

Earlier, field survey teams wrote down observations in a narrative style. The communication modes used for conveying information were post, telegrams, and telex, and the flow of information was irregular, much delayed and the statistics were often incomplete or vague. The use of facsimile machines followed by the introduction and adoption of fax, computer, desktop publishing, Internet, etc. makes it possible to know the locust situation in any part of the world and makes it available in database form. The e-mail service fastens the sending and receiving of data, bulletins, and additional information beneficial for locust control. By the year 2000, it became possible to exchange and communicate data and information through Internet services and e-mail among almost all nations (Cressman and Locust, 2016). Food and Agricultural Organization (FAO) signaling and forecasting updates are consistently updated on the Internet and transmitted by e-mail. The advancements in communication modes help to spread information more rapidly and strengthen the evaluation of the situation and ecological conditions of locusts that in turn bring improvements in forecasting.

4.2.2 GEOSPATIAL DATA

The Geographic Information System (GIS) is a computerized system involving the storage, display, manipulation, and analysis of geographic data related to the oviposition, density of hopper bands or swarms, area, temperature, precipitation, greenness, etc. GIS software is improving day by day in number and functions. The data pertaining to locusts, their environment, especially rains from different parts of the world are integrated into a computer system and are used for forecasting purposes. GIS is perhaps the most suitable technology to assist forecasting and research on locusts. *S. gregaria* has a sophisticated GIS version (SWARMS, *Schistocerca* Warning Management System) that operates from a workstation under FAO in Rome covering the whole breeding area. Its simplified versions work from portable computers from different countries (RAMSES, Reconnaissance and Management System for the Environment of *Schistocerca*) (Magor, 1993; Cressman, 1997; Magor and Pender, 1997; Rosenberg, 2000). A monitoring system, Emergency Prevention System (EMPRES, desert locust component) established by FAO was implemented in Western and Northern Africa in 2006 for an early survey and preventive control of *S. gregaria* populations in their reproduction zones (Bonnal et al., 2010). In the year 2014, an open-source platform independent version evolved with the use of Open Jump GIS and Postgres spatial database (Cressman and Locust, 2016). The Geographical Positioning System (GPS) technology is used to record the geographic coordinates during the collection of field data. These coordinates are then exported to image processing or GIS software. This method vanishes mistakes related to transcribing field notes and annotated coordinates in maps. Various organizations, the world overemploy this technique to gather regular information quickly and decrease the time to update field statistics (Latchininsky and Sivanpillai, 2010).

4.2.3 REMOTE SENSING

The ground networks are less efficient because of the wide geographic scale of locust infestations, not yielding enough data for correct evaluation of population dynamics, and being unable to assist in deciding the exact time for the application of control measures. Remote sensing technology brings substantial improvements in more precise interpretation. The major objectives to use remote sensing are (1) to forecast the zones which are at higher risk of locust infestation and (2) to appraise the level and seriousness of destruction. The remote sensing data are important to apply appropriate control measures in time. Coupled with field data and geospatial data (GIS and GPS technology), remote sensing proves as a more effective approach in locust management programs. Remote sensing techniques are in use since 1970s for studies pertaining to locust outbreaks. Satellite remote sensing was used for the first time to detect the locust breeding sites in 1973 (Pedgley, 1973). Since then, remote sensing becomes an important tool for the management of locust outbreaks to monitor locust ecology and population (Hielkema, 1977; Hielkema, 1980), to identify and monitor habitats depending upon the vegetation greenness (Mcculloch and Hunter, 1983), to examine ecological conditions of *S. gregaria* recession areas (Tucker et al., 1985), to locate regions worthy for breeding and egg laying by observing the variations in vegetation conditions (Bryceson, 1984, 1989; Bryceson and Wright, 1986), to confine areas of rain and developing vegetation supporting breeding, thereby directing the field survey teams (Cherlet and Gregorio, 1991; Cherlet and Schistocerca, 1993; Voss and Dreiser, 1997), to study the incident and magnitude of locust outbreak in reed marshes and Normalized Difference Vegetation Index (NDVI), to examine the most optimum zones for locust hatching (Shi et al., 2003), to map and categorize the zones infested by locusts into light, moderate, and heavily damaged classes (Ji et al., 2002), and to monitor damage caused by locusts (Ma et al., 2005). Landsat Multispectral Scanner System (MSS), Moderate Resolution Imaging Spectroradiometer (MODIS), and National Oceanic and Atmospheric Administration-Advanced Very-High-Resolution Radiometer (NOAA-AVHRR) sensors are mostly utilized for remote sensing. These have a spatial resolution of 78 m, 250 m, and 1100 m, respectively and temporal resolution of 18 days, 1–2 days, and 1 day with 0.6, 6.25, and 121 ha of each pixel area, respectively. Landsat MSS has high spatial but low temporal resolution and is incapable of efficient monitoring. MODIS and NOAA-AVHRR have higher temporal resolutions but too low spatial resolutions fail to exactly describe the locust destruction at the local level. Landsat Enhanced Thematic Mapper (ETM)+/Thematic Mapper (TM) have higher spatial 30 m and temporal 16-day resolutions compared with Landsat MSS. Therefore, Landsat ETM+/TM images are available to more precisely determine the geographic magnitude of locust destruction at the local level as the period from first hoppers to adult locusts is nearly a month. Multi-temporal Landsat ETM+data are used to analyze the extent and severity of *L. migratoria manilensis* plague (Tian et al., 2008). Landsat, MODIS and Satellite Personal Tracker (SPOT) satellites are also used for *L. migratoria migratoria* (Sivanpillai et al., 2006; Navratil, 2007; Sivanpillai and Latchininsky, 2007), and MODIS for *L. migratoria manilensis* to monitor and analyze pre- and post-locust outbreak situation

(Ji et al., 2004; Liu et al., 2008). High-resolution SPOT satellite statistics is helpful to track down *N. septemfasciata* habitats in Madagascar (Franc, 2007). Indian Remote Sensing (IRS) P6 Satellite-Advanced Wide Field Sensor (AWiFS) data are employed to trace the habitats of *C. italicus* in NE Kazakhstan (Sivanpillai et al., 2009). In the Western US, an advisory system, CARMA is used for the management of grasshopper infestations since 1996 (http://carma.unk.edu/; Hastings et al., 2009). The FAO started using 250 m resolution MODIS imagery containing a 16-day composite image of NDVI values (FAO 2009; Pekel et al., 2010). Piou et al. (2013) assessed the link between the historical prospection statistics and remotely sensed NDVI values given every 16 days and at 250 m spatial resolution (MOD13Q1 from MODIS satellite) for the preventive control of *S. gregaria*. The Desert Locust Information Service (DLIS) of the FAO (UN) utilizes the particulars from satellites for early warning of locust outbreaks (http://www.fao.org/ag/locusts/en/activ/DLIS/satel/index.html). DLIS in collaboration with other universities and institutes, for example, International Research Institute (IRI) for Climate and Society Columbia, the Italian Institute of Biometeorology (IBIMET), the European Commission Joint Research Centre (JRC), NASA's World Wind Project and the Catholic University of Louvain (Belgium) is actively engaged in enhancing the use of remote sensing imagery in observing and early warning of *S. gregaria* (Latchininsky and Sivanpillai, 2010). The latest products with greater resolution, such as the European Space Agency's PROBA-V and Sentinel 3 with 100 and 30 m resolutions, respectively are more effective to find out the greenness in *S. gregaria* habitats (Cressman and Locust, 2016).

4.3 PHYSICAL METHODS

Through the incorporation of modern scientific tools, the conventional physical locust control methods undergo various improvements. Physical traps accompanied by optical and mechanical collection technologies prove beneficial in reducing the plague in highly infested zones. The more effective and sustainable approach is the incorporation of precision Bayesian prediction modeling (Meynard et al., 2020; Walsh, 1986; Xu et al., 2005; Zhiwei et al., 2007; Maleki and Khorram, 2010). As the locusts have phototaxy behavior, specific light and sound wave lengths act as stimulants enhancing their trapping because of intrusion in sight and audio recognition by interacting with glutamic acid and dopamine neurophysiology (Ripley and Ewer, 1951; Michelsen, 1968; Daniel and Haig, 1990; Ostfeld and Keesing, 2004; Wang and Zhou, 2014). Sometimes, the availability of food/host affects the locust density that in turn influences the locust outbreaks. Heavy livestock grazing results in more *Oedaleus asiaticus* (a nonmodel economically destructive locust) outbreaks. *Oedaleus asiaticus* feeds on plants having less N content and artificial diets of low protein content but rich in carbohydrates. Since the N content of plants is the lowest in fields facing excessive grazing and have extremely low N content, possibly because of increased soil erosion. Hence, locust abundance is the highest. Thus, one strategy to control this locust outbreak is to keep livestock grazing at minimum (Cease et al., 2012).

4.4 CHEMICAL CONTROL

Earlier dieldrin and other organochlorine insecticides were used against locusts. Later on, these were banned in many countries as these exerted adverse environmental and health effects. Other replacement insecticides, such as carbaryl, carbofuran, dimethoate, malathion, monocrotophos, phenthoate were much less persistent and require frequent application in blanket treatments and in greater volumes. Although, these were less toxic than dieldrin, their effects on the environment and human health are more hazardous (Vincent et al., 2007; Peshin, 2014; Usmanov and Gapparov, 2020). The toxic and bioactive substances present in the pesticides can interact with the nontarget organisms, influence soil productivity and ecosystem stability, possess longer persistence, and disturb soil microflora and soil health (nutrient composition, organic carbon, pH, water content, enzymes, etc.) (Prashar et al., 2013). These molecules may enter water bodies and cause toxicity in aquatic fauna (Hough, 2014). These toxic compounds can contaminate food items and can lead to endocrine disruption, asthma, cancer, diabetes, cognitive effects, or male reproductive disorders (Baker et al., 2002; Kim et al., 2017; Song et al., 2017). With time, improvements are seen in chemical locust methods due to the discovery of novel and more effective insecticides, insecticidal formulations, and various insecticide application techniques. The nitrogen (N) fertilizers decrease the survival and performance of the Senegalese locust (*Oedaleus senegalensis*) as this locust exhibits higher preference and performance in low protein and high carbohydrate diet. N fertilizer enhances the protein content of the host plant and results in higher protein-to-carbohydrate ratio that lowers the survival and reproduction in females (Gall et al., 2020). To reduce the amount of chemical insecticides used against locusts, the insecticides are sometimes applied in combination with pheromones and some other insect growth regulators (Bashir et al., 2016). Locust swarms are sprayed with insecticides mainly when the insects are settled or in flight. Introduction of new pesticide application techniques and tools makes it possible to apply insecticides more precisely and to a greater area in a shorter time. The application of insecticides is often made as per the Ultra Low Volume (ULV) treatment technique. This strategy utilizes a lot littler volumes of shower fluid known as ULV splashing. It is the most productive and commonly used strategy utilizing usually 0.5–5 L of shower fluid per hectare. Unique sprayers are required for ULV splashing. These sprayers can be versatile, vehicle-mounted or airplane-mounted. Currently, the locust invasions are sprayed with ULV definitions of contact pesticides by using the sprayers, namely, microULVA, ULVAmast, and Micronair AU 8115. Most recently, drones are the most effective means for spraying insecticide (Jayaprakash et al., 2016). The ULV technique coupled with GPS, GIS, and remote sensing proves more beneficial (Moharana et al., 2020). The FAO given GPS technology, especially of Differential Global Positioning System (DGPS), guided by satellite brings revolution in locust control by (1) securing navigation, mainly in desert areas where *S. gregaria* occurs frequently, (2) upgrading the survey by recording the exact coordinates of target locusts and transferring them to spray aircraft, and (3) significantly upgrading precision spraying. Precision to within a few meters can be attained in finding the targets and directing the pilot spray aircraft (the GPS directs the passage of the plane in the desired route to within a few meters, at the desired speed and constant width between each passage without using ground beacons). It keeps checking on the insecticide rate automatically and has significant precision in the doses spread over a hectare (Dobson, 1999; Ottesen et al., 1999). The GIS-based analysis helps to trace the movement of locusts and hence informs about the invasion route for the application of pesticide, for example, the recent movement of *S. gregaria* and *L. migratoria* toward the Indian subcontinent. Computational chemistry reveals the effectiveness of agricultural insecticides against locusts depending on their binding affinity for the locust's survival proteins. The phylogenetic evaluation displays the possibility of effectiveness and safety of the recommended pesticides against locust species (Banik et al., 2020). Chemical control reduces the contaminated surface area gradually as well as the number, size, and density of swarms (Lecoq, 2000). Additionally, it stops the locust plague in a shorter period as compared with the previous duration of plague. In spite of various merits of chemical locust control, alternative approaches are still in great demand.

4.5 BIOLOGICAL CONTROL

The chemical insecticides contaminate groundwater and are hazardous for the natural enemies of locusts. The growing resistance of agricultural pests against chemical insecticides makes it essential to find alternative biological control agents (Rajput et al., 2020). Biological control agents can be categorized into (1) predators, (2) microbes, and (3) biochemicals.

4.5.1 PREDATORS

Human diet and animal feed comprised insects aid in alleviating poverty (FAO, 2010). In poultry, insects are used as feed in place of fishmeal and soybean meal (Khan, 2018). *S. gregaria*, *L. migratoria*, *Locustana pardalina* and *Nomadacris septemfasciata* red locust are common locust meal for poultry. Locusts are fed live or after drying and in powder form (broilers). In some cases, the dried locusts are served after boiling (Khusro et al., 2012). Locusts are a good protein-rich diet for fishes and shrimps. The conversion of lowland areas into fishponds and shrimp farming is a sustainable measure to control locust outbreaks through predation (Wang et al., 2011). The release of locust predators in the field, such as wasp larvae, mites, spiders, and birds feeding on larvae and developing locusts are known to control locust outbreak up to 90% (Goodman, 1993; Karen et al., 2006; Arroyo et al., 2019; Sharmila, 2020; Anonymous. Locust Alarm, 2020).

4.5.2 MICROBES

There are two ways to use microbes as biological control agents: (1) classical use where the pathogen is not commercially produced or conserved as a naturally occurring pathogen in the agro-ecosystem, (2) augmentative where the pathogen is produced on commercial scale. Nearly, 50 entomopathogenic viruses, bacteria, fungi, and nematodes are produced commercially and used augmentatively as microbial pesticides (Lacey et al., 2015). These are easy to operate, propagate, formulate, and exhibit great levels of pest control efficacy (Glare et al., 2012). Various microbes are used against locusts.

4.5.2.1 LOCUSTICIDAL FUNGI

There are approximately more than 1000 species of fungi belonging to nearly 100 genera known as entomopathogenic fungi capable of infecting insects of various orders and acting as mycopesticides (Shah et al., 2009; Vega et al., 2012). Mycopesticides are products made from live propagules of fungi supplemented with a nonreactive substance or an adjuvant facilitating their easy operating, use, and enhance their efficacy (García de León and Mier, 2010). These mycopesticides are available in several forms, such as powder, dispersible and water-soluble granulate, and aqueous suspensions. The procedure of use, composition, and environmental factors play pivotal role in efficiency, persistence, and spatial distribution of propagules of fungi. Mycopesticides are a part of the list of products recommended for locust control (FAO, 1999). Green Muscle® and Green Guard® from (*Metarhizium acridum*), Mycolar B® from *Beauveria bassiana* and Mycolar M® from *Metarhizium anisopliae* are the commercial mycopesticides against locusts (Faria and Wraight, 2007; Lednev et al., 2018). The entomopathogenic species are spread over Ascomycota, Basidiomycota, Deuteromycota, Mastigomycota, and Zygomycota (Burges and Hussey, 1971; Whitten and Oakeshott, 1991; Starnes et al., 1993). A majority of the entomopathogenic species belong to the order Hypocreales (Ascomycota) and have a wide host range while Entomophthoromycota (Zygomycota) species infect specific hosts. These fungi can be obligate or facultative arthropod pathogens. The entomopathogenic fungi survive as saprophytes and endophytes of plants (Goettel and Johnson, 1997). *Entomophaga grylli* is a species complex with many pathotypes with antilocust action against *L. migratoria* and *D. maroccanus*. The anamorphic fungi, *Beauveria*, *Metarhizium*, *Isaria*, *Hirsutella*, *Lecanicillium*, *Aspergillus,* etc. have been reported in locusts (Lednev et al., 2020). *Aspergillus flavus* var. *oryzae* has also been observed to act against the migratory locust *L. migratoria* (Zhang et al., 2015). *Metarhizium* spp. (Hyphomycetes) are distributed from the arctic to the tropics across the globe and they reproduce by conidia formation (Bukhari et al., 2010). *M. acridum* is a specialist attacking Orthoptera (grasshoppers and locusts) and *M. anisopliae* is a generalist attacking Coleoptera and other insects, such as mosquitoes, ticks, termites (Wang et al., 2012; Hu et al., 2014). *Metarhizium* spp. are found in both cultivated and natural soils and have been employed extensively for pest control due to their short host range, safety, eco-friendly nature and easy large-scale production. The infection-causing units are conidia, utilized as mycopesticides. These species attack at different stages during the development of an insect. There are three ways of locust infection through conidia: (1) direct contact, (2) by secondary pick up of conidia from treated vegetation, and (3) horizontal transmission from conidia-contaminated cadaver. The lifecycle of *Metarhizium* spp. on a locust involves different stages as shown in Figure 4.1.

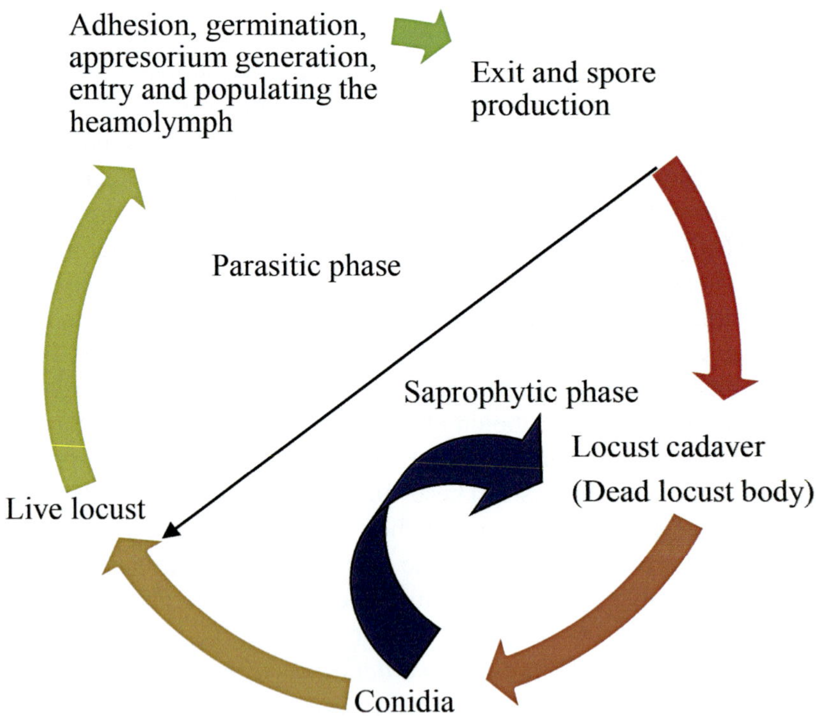

FIGURE 4.1 Life cycle of a Metarhizium sp. on a locust.

The conidium adheres to the host body surface, germinates to form a tube with an appresorium at the tip. The appresorium due to mechanical force (hyphal pressure) or enzymatic action (lipases, proteases, and chitinases) enters the insect cuticle. A micropore is formed that approaches the hemocoel. The fungus utilizes the host nutrients for growth and development, reproduces, and increases in biomass, attacks the malpighian tubules and nervous system, saturates the immune system, and ultimately leads to the death of the insect. The fungus then exits the parasitic phase and enters the saprophytic phase, feeding on the cadaver (dead insect body), producing conidia/asexual spores for further infection (Aw and Hue, 2017). During sporulation, *M. anisopliae* hyphae come out of the host cuticle and constitute a dense mycelial network bearing green spores on the dead body of insects while *M. acridum* spores on locusts and grasshoppers cadavers appear red (Seyoum and Negash, 2007; Ocampo and Caoili, 2013; Gabarty et al., 2014). *M. anisopliae* and *M. acridum* conidia are utilized as potential ingredients of mycoinsecticides respectively against pests (Faria and Wraight, 2007). The younger nymphal instars are more sensitive than the elder ones. The most probable reason behind this is the upregulation of expression level of heat shock proteins (Hsps) in gregarious locusts than those in solitary ones (Wang et al., 2007). Therefore, younger nymphal instars are preferably targeted to prevent the development of elder nymphs into adults which could multiply to increase the number of locusts. *Metarhizium* mycopesticide causes 70–90% mortality of locusts in 14–20 days without any noticeable effect on nontarget fauna (Kooyman et al., 1997; Lomer et al., 2001). Green Muscle® from *M. acridum* (formerly named as *M. anisopliae* var. *acridum*) has potential against the sexual locust nymphal instars, *Dericorys albidula. M. acridum* isolate, IMI 330,189 formulation is significantly effective against the second to fifth nymphal instars of *D. albidula* (Valizadeh et al., 2011). The infection with *M. anisopliae* CQMa421 strain causes the mortality of *L. migratoria manilensis* through interaction with the locust' immune responses. *M. anisopliae* treatment lowers the activities of superoxide dismutase (SOD) and prophenoloxidase (ProPO). Moreover, the function of *L. migratoria* genes (defensin, spaetzle, and attacin) concerned with the immune system differs upon 1–4 days of application of *M. anisopliae* (Jiang et al., 2020). *M. anisopliae* exhibits enhancement in catalase–peroxidase activity at the time of germination and growth. These enzymes function to convert hydrogen peroxide into water and oxygen. Transgenic strains of *M.*

anisopliae that overexpress the CAT1 gene have been observed with double catalase activity than that of the wild-type strain. The enhancement in the activity of catalase improves resistance to exogenous hydrogen peroxide and decreases the germination time (Hernandez et al., 2010). Several factors contribute toward the efficacy of *Metarhizium* spp. as mycopesticides. The use of chemical fungicides, Antracol 70 WG, Dithane Neo Tec 75 WG, and Score 250 EC formulation create hindrance in the growth of *M. anisopliae* conidia. The culture medium supplemented with soil extracts containing fungicides prevents the conidial growth. Extract from organic soil having fungicides strongly inhibits *M. anisopliae* colonies as compared with the extract from sandy soil (Tkaczuk et al., 2013). The type of culture medium influences the characteristics and pathogenicity of *Metarhizium* spp. conidia. Sabouraud dextrose agar with 1% yeast extract results in the production of the thickest *M. anisopliae* conidia and the maximum colony growth with the local isolate (PikKheng et al., 2009). The potato dextrose agar along with 1 g/L yeast extract produces rapidly germinating conidia with more UV-B tolerance than those produced using insect cuticle. This is advantageous for the large-scale production of more virulent conidia (Rangel et al., 2005). The best virulence temperature range for *M. acridum* is 28–33°C. At 28°C, maximum conidial germination, the longest mycelial formation, and the highest conidial production occur (Nyam et al., 2015). Several studies indicate that locusts having the potential forthermoregulation can bask in the sun upon infection with *M. acridum*. Basking elevates the temperature of the body to approximately 35°C (37–40°C in summer), and hence protects the locusts from infection (Long and Hunter, 2005; Mullié and Guèye, 2010). Different strains of *M. acridum* differ in virulence due to genetic differences or because of the difference in the climate of different ecogeographic regions. The regions with long daylight and high temperature need more virulent *M. acridum* strains as the locusts are capable of elevating their body temperatures and vice versa (Niassy et al., 2011). The conidia sprayed onto soil exhibit greater viability (1 year and 4 months) than those sprayed onto the vegetation (8 months). The virulence of *M. acridum* is not affected by the vegetation thickness. The number of *M. acridum* conidia on the soil shows negative correlation with relative humidity and positive correlation with wind velocity. More research is needed to find the most appropriate application method and the minimum conidial dose resulting in the highest mortality (Guerrero-Guerra et al., 2013). For *M. acridum*, the conidial emulsion formulation is a much better option in comparison to oil formulation, because the former displays greater persistence in the environment particularly under low humidity besides having equivalent tolerance toward heat and UV-B tolerance to oil formulation (Peng and Xia, 2011). The oil in emulsion formulation safeguards conidia from high temperature and sun rays (Chouvenc et al., 2011). However, dry *M. acridum* conidia spore formulation has greater shelf life than oil formulation (Diedhiou et al., 2014). Mycopesticides being slow in action during mild to cool conditions, significant survival percentage when applied against dense infestations, sporadic nature of outbreaks, and high price limit their use. However, they may play a role in integrated locust management strategy alongside classic insecticides (Hunter, 2005). Recent research focuses on improving the virulence of *Metarhizium* spp. through biotechnology and molecular studies. The cuticle-degrading enzymes, chitinase, protease, and lipase participate in the penetration process and hence can affect the virulence of the fungal strains (Bai et al., 2012). The addition of olive oil, surfactants (SDS and Tween 80), magnesium to the basal medium, pH maintenance (5.7), and temperature (32°C) enhances the production and activity of lipases extracted from *M. anisopliae* (Ali et al., 2009). This in turn could improve the virulence of the fungus against insects. Molecular studies enhance the virulence of *Metarhizium* spp. either by overexpression of pathogenesis-related genes or by genetic manipulations. Transferring the esterase gene (MestI) from *M. robertsii* to *M. acridum* expands the host range of the latter species (Wang et al., 2011). The transfer of scorpion toxin (BjαIT) gene into *M. acridum* increases pathogenicity of the fungus against *L. migratoria manilensis* (Peng and Xia, 2015). The incorporation of a *Leiurus quinquestriatus hebraeus* (LqhIT2) scorpion venom gene into *M. acridum* genome enhances fungal virulence. The resulting *M. acridum* strain grows rapidly and decreases the half-lethal time (LT_{50}) and half-lethal concentration (LC_{50}) by 30.3% and 22.6 times, respectively in comparison to the wild-type. The resulting neurotoxin does not affect the cuticle piercing and germination but suppresses the locust immunity, accelerates conidial growth in the hemolymph (Peng and Xia, 2014). The ATM1 gene overexpression on *M. acridum* increases acid trehalase production. This acid trehalase breaks trehalose in the locust hemolymph into glucose, generating greater energy for *M. acridum*. This in turn reduces the growth and colonization time of *M. acridum* and enhances the virulence of *M. acridum* (Peng et al., 2015). The co-inoculation of strains with AaIT1 (a sodium channel blocker) and hybrid-toxin (a blocker of both K and Ca channels) expressions, showing additive results, improves the virulence of *M. acridum* for acridids by 11.5-fold decrease in LC_{50}, 43% decrease in time to kill (LT_{50}) and 78% decrease in food consumption without causing disease in non-acridids (Fang et al., 2014). Silencing the *L. migratoria* inhibitor apoptosis protein 1 (LmIAP1) gene causes direct mortality and enhances host susceptibility to *M. acridum* by reducing the immunity and modulating the gut microbiome (Zhang et al., 2019). Genetic manipulations enhance the virulence of the fungus. However, further studies are required to investigate (1) how to inhibit the decrease in the production of conidia, and (2) to appraise the effects of the genetically modified fungus on environment and nontarget fauna. Sometimes, the mixed infections of entomopathogenic fungal strains with other locust pathogens prove to be more effective as locust control agents. *Beauveria bassiana* (EABb 90/2-Dm) and *M. acridum* (IMI 330, 189) strains when used together display synergistic action on the longevity, eating, and multiplication of *D. maroccanus* reducing the feeding and fecundity (Valverde-Garcia et al., 2019). *B. bassiana* infection followed by the microsporidium, *Paranosema locustae* infection shows greater mortality and changes in the gut microflora than the simultaneous infection and was more effective than either treatment alone against *L. migratoria* (Tan et al., 2020). Because of the adverse impacts of chemical insecticides on the environment, humans, and other nontarget organisms, it is necessary to accelerate the examination of mycopesticides before commercialization (Brunner-Mendoza et al., 2019). Many regulatory authorities worldwide like the International Organisation for Biological Control (IOBC) and the Environmental Protection Agency (EPA) promote and develop protocols for registering the microbes employed as biopesticides. As per the proposal of the World Health Organization (1981), the following data and tests are considered for the registration of a biopesticide: product examination, residue testing, toxicogenic effects, the impacts on nontarget organisms and environment, and effectiveness and functionality (Siegel and Lacey, 1997). Currently, for registration of a biopesticide, data pertaining to the identity (scientific name) and components of the microbial pesticide (inert ingredients), physicochemical characteristics (color and pH), biological properties, toxicity profile, ecotoxicological data, and stability profile is made compulsory (Sanitarios and (COFEPRIS). X Reglamento En Material De Registros, 2005). Many tests concerning the biosafety of fungi and other microbes used as biocontrol agents have been conducted to get information related to their toxic characteristics, environmental safety such as the effect on the stability of the ecosystem because of their persistent nature (Zimmermann, 1993). Additionally, the entry of exotic strains should be questioned (Lockwood, 1993). To appraise the precarious health effects on humans, in vivo studies have been performed on mice and rabbits using several Mexican fungal strains (Mier et al., 2005; Brunner-Mendoza et al., 2017). The virulence and toxigenic nature of *M. anisopliae* (Toriello et al., 1999, 2006) and the acute effects of gastric exposure of *M. acridum* (Toriello et al., 2009) in mice have been evaluated. These investigations in general display low health hazards of *Metarhizium* strains in mammals. *Metarhizium* is reported to be allergenic but with no major hazardous impacts on the production staff or users (Zimmermann, 2007). However, a few studies reveal the adverse health impacts, such as scarce human infections caused by a few strains of *Metarhizium* spp. (García et al., 1997; Jani et al., 2001; Motley et al., 2011; Showail et al., 2017; Goodman et al., 2018; Amiel et al., 2008; Eguchi et al., 2015; Burgner et al., 1998; Revankar et al., 1999). Therefore, research related to the biosafety of biopesticides must continue as we are operating living organisms. The impact on nontarget fauna including humans should be assessed while utilizing living organisms for locust management.

4.5.2.2 LOCUSTICIDAL BACTERIA

Soil bacteria play a pivotal role in crop production. Some are nitrogen fixers while others are important in the cycling of fertilizers. Certain forms of bacteria protect plants against pests through the production of antibiotics or toxins (Lacey et al., 2015). Only a few bacterial strains act as parasites of locusts (such as *Enterobacter cloacae*, *Serratia marcescens*, *S. entomophila*, and *Pseudomonas aeruginosa*) (Lomer et al., 2001; Mashtoly et al., 2019). *Yersinia* sp. (MH96) is a promising biopesticide active against *L. migratoria* early instars (Hurst and Glare, 2006; Mcneill et al., 2008). The behavior of several *Yersinia* spp. changes with temperature and this thing could influence the bacterial efficacy in the field (Chester and Stotzky, 1976; Logue et al., 2000). The gram-negative bacterium, *Serratia marcescens* extracted from *S. gregaria* in Kenya, is now a well-established pathogen of grasshopper and locusts (Tao et al., 2006). An insecticidal protein (Pr596) from *S. marcescens* HR-3 is a metalloprotease with antilocust action (Tao et al., 2007). *Bacillus thuringiensis* (Bt) produces Bt endotoxin (Crystal 7A) that is lethal against *L. migratoria manilensis* (Song et al., 2008; Wu et al., 2011). The bacterial strains, namely, *Pseudomonas* sp. strain B3 (HF911369), *Pseudomonas* sp. strain B4 (HF911366) *Enterobacter* sp. strain B6 (HF911368), and *Bacillus* sp. strain B5 (HF911367) exhibit locusticidal potential for *L. migratoria* and exhibit lethality of 100, 98, 71, and 65%, respectively (Oulebsir-Mohandkaci et al., 2015). Synchronous infection with *Pseudomonas* sp. bacteria and fungi (*B. bassiana, M. anisopliae*) causes greater mortality in a shorter time reducing the LT_{50} to about 3 days as compared with monoinfections by bacteria and fungi (Lednev et al., 2008).

4.5.2.3 LOCUSTICIDAL MICROSPORIDIA, NEMATODES, AND VIRUSES

Unlike fungi and bacteria, only a few examples of microsporidia, nematodes, and viruses with antilocust efficacy are known till date. Microsporidia are obligate unicellular parasites reproducing in other living hosts. These were previously considered protozoans but molecular studies have classified microsporidia inside the Kingdom Fungi. Approximately, 186 genera of microsporidia act as insect pathogens (Keeling, 2009). The microsporidium, *Paranosema locustae* affects the morphological phase transformation of *L. migratoria manilensis* (Fu et al., 2010). It prevents the locust swarm behavior by inhibiting the aggregation of solitary *L. migratoria manilensis* locust and inducing the gregarious locusts to shift back to solitary phase. *P. locustae* acidifies the hindgut, alters the immune response, inhibits hindgut bacteria to grow that produce aggregation pheromones. This in turn decreases the production of neurotransmitter (serotonin) that starts the gregarious behavior. The parasite-infected locusts produce less serotonin, reducing aggregation of the solitary locusts. On the other hand, the infected gregarious locusts shift back to solitary phase as the pathogen reduces the production of neurotransmitter, dopamine necessary to maintain gregarization (Shi et al., 2014). *P. locustae* shows synergistic action with *M. anisopliae* var. *acridum* against *S. gregaria.* The nymphs treated with both the pathogens died sooner as compared with the nymphs treated with either of the pathogen (Tounou et al., 2008). The locusticidal nematodes *Mermis nigrescens* and *Agamermis decaudata*

(Mermithidae) are endoparasites of locusts. The nematodes larvae reach the hemocoel, feed on the nutrients, and exit the host for the completion of their life cycle in the soil. As the nematodes emerge out of the locust body, the host dies (Baker and Capinera, 1997). Nematode, *Steinernema carpocapsae* applied to the fifth instar nymphs exerted a concentration and time-dependent impact on the mortality of *L. migratoria* and caused 100% mortality at the third day of treatment. The nematode broke the host immunity through changes in the enzyme activities (phenoloxidase, peroxidase, α and β estrases and glutathione-S-transferase) (Wahed et al., 2018). Keeping in view of the entomopathogenic potential of nematodes, efforts are in progress to produce them on large scale (Cortés-Martínez and Chavarría-Hernández, 2020). Among viruses, the DNA-containing virus, entomopoxvirus has been reported to infect locusts. Till date, entomopoxviruses are reported from 15 species of grasshoppers and locusts (Erlandson and Streett, 1997). The virus (SINPV-type B) responsible for nuclear polyhedrosis in *Spodoptera littoralis* could be cross-transmitted perorally to and from two locusts species viz. *L. migratoria migratorioides* and *S. gregaria*. The virus caused the disease "dark cheeks" leading to lethality of the locusts (Bensimon et al., 1987). The cytoplasmic polyhedrosis virus enhanced the mortality of the migratory locust by bacteriosis due to *Enterobacter cloacae* (Lednev et al., 2020).

4.5.3 BIOCHEMICALS WITH ANTILOCUST EFFICACY

4.5.3.1 BOTANICALS

Research has been carried out and is in progress on pest control through the application of botanicals (Isman, 2020). Sometimes, plants with antilocust potential are fed directly to locusts. *Nerium oleander* (Apocynaceae) leaves fed as staple food exhibit toxic effects on development and food intake leading to mortality of *S. gregaria* fourth instar larvae. This could possibly be the contribution of the toxic secondary compounds present in the plant leaves (Bagari et al., 2013). The extract of *N. Oleander* leaves used as staple food inhibits the ovarian development of *S. gregaria* females and enhances the mortality (Bagari et al., 2015). *Azadirachta indica* seed extract shows toxicity against *S. gregaria*. The toxicity of the extract increases about ten-fold with the posttreatment temperature elevation from

22°C to 40°C (Kabaru and Mwangi, 2000). The root extracts of *Mucuna pruriens* show antilocust potential against *L. migratoria* and *S. gregaria* (Abdalla, 2004). The extracts of *Fagonia bruguieri* remarkably alters the functioning of acetylcholine esterase in the hemolymph of *S. gregaria* nymphs and adults (Ghoneim et al., 2012). The extracts of *Nigella sativa* seeds exert toxic impacts on several parameters of the adult performance and phase transition of *S. gregaria* (Ghoneim et al., 2015). The rags treated with the foliar extracts of *Cleome arabica* and fed to *S. gregaria* exhibit mortality of 76.67% and 86.67% in male and female larvae respectively and 100% mortality of the male and female imagoes. These biotoxic effects result from the impairment and damage of the digestive tract by *C. arabica* extracts (Kemassi et al., 1775). The aqueous extracts of *Schinus molle* greatly decrease the L_5 population of *L. migratoria*. The toxicity and hence the mortality results from the secondary molecules (such as limonène, α phéllandrène, timol, citronellylacetate, and β-cariophyllene and cis-menth-2 and-1-ol y trans-piperitol) present in the plant leaves (Chilali and Benrima, 2018). *Melia volkensii* is a dryland tree species native to East Africa, the extract of this plant species inhibits growth and exhibit an anti-feedant effect on *S. gregaria* (Jaoko et al., 2020). The antilocust efficacy of plant extracts can be attributed to a number of antilocust metabolites present in them. Lycorine alkaloid isolated from the bulbs of *Hymenocallis littoralis* when sprayed (at concentration of 0.05%) on cabbage leaves inhibits the feeding of *S. gregaria* (Singh and Pant, 1980). *Calotropis procera*, *Zygophyllum gaetulum*, and *Peganum harmala* alkaloids cause weight loss, prevention or delay in sexual maturity, decrease in female fecundity and hatching rate, and enhance the mortality of *S. gregaria* (Abbassi et al., 2003). The treatment with the sesquiterpenoid, farnesol reduces food consumption, changes nutritional indices, and disturbs the digestive enzymes (protease, invertase, amylase, trehalase, and chitinase) of *S. gregaria*. The treated insects oviposited fewer eggs (Awad et al., 2013). The synergistic action of the botanicals enhances their locusticidal potential. Different solvent extracts of *A. indica* and *N. sativa* disturb the survival, inhibit the growth and development, and intervene with the metamorphosis of *S. gregaria* (Sh et al., 2013). A zooecdysone (crustecdysone) obtained from *Podocarpus gracilior* leaves or a plant growth regulator (Alar® 85) injected into newly ecdysed, gregarious, fourth instar nymphs of *S. gregaria* induce solitarious character and only Alar® 85 causes adult sterility (El-Ibrashy et al., 1976). The aqueous extracts of *S. gregaria*, the

Bombay locust (*Nomadacris succinct*) also known as *Patanga succincta* and the migratory locust (*L. migratoria*) frass as well as aqueous extracts of leaves of *Bromus catharticus*, *Dactylis glomerata*, *Sorghum bicolor*, Japanese *Brassica rapa* var. *perviridis*, *Lactuca sativa* var. *longifolia*, and *Brassica oleracea* var. *capitata* exhibit inhibitory impacts on the oviposition and embryonic development of *S. gregaria* adults (Mansour et al., 2015). The plant essential oils also exhibit antilocust efficacy (Tanaka et al., 2019). The essential oils of *Artemisia herba-alba* enriched with chrysanthenone, camphor, α-thujone, α-pinene, and β-thujone exhibit antilocust activity against *Euchorthippus albolineatus* locusts. The lethal time (LT_{50}) is nearly 1.67 and 1.45 days for males and females, respectively (Zaim et al., 2012). Caraway, orange peel and wintergreen oils combined together in linseed oil/bicarbonate emulsion exhibit toxicity against the gregarious *S. gregaria* and *L. migratoria*. After single spray treatment, the synergistic action of this toxic formulation causes a mean mortality rate of 80% and 100% of *S. gregaria* and *L. Migratoria*, respectively within 24 h (Abdelatti and Hartbauer, 2020).

4.5.3.2 PHEROMONES

Pheromone traps could detect the transient stage of locust population growth and hence aid in forecasting the outbreak of upsurges. These may act as locust control agents by making locusts transient and more sensitive to predators or by enhancing their sensitivity to insecticides even at sublethal doses (Hassanali and Bashir, 1999). Phenylacetonitrile pheromone of *S. gregaria* causes mortality of the nymphs. The same pheromone along with sublethal doses of fipronil, malathion, and carbosulfan enhances the insecticidal efficiencies by making the nymphs more sensitive toward these pesticides (Bashir et al., 2016). Similarly, phenylacetonitrile along with *M. anisopliae* var. *acridum* exhibits synergistic action against *S. gregaria* fifth instar nymphs and makes them hypersensitive and disoriented (Abdellaoui et al., 2020). Pheromones act as mediators in the exchange of information between conspecifics and shape distinct behaviors like reproduction (mate finding and courtship) and aggregation (Hassanali et al., 2005; Hansson and Stensmyr, 2011). The formation of swarm needs large-scale reproduction and aggregation. Pheromones for reproduction and aggregation behavior are detected via the receptors of the olfactory system. Several pheromone receptors have been identified from *S. gregaria* olfactory system mediating responses to pheromones controlling reproduction and aggregation (Pregitzer et al., 2017). As an adaptation to mating competition in gragarious condition, males of *S. piceifrons* produce volatile sex pheromones viz. phenethyl alcohol, 2-phenylethanol, (*Z*)-3-nonen-1-ol (3-Nol), and (*Z*)-2-octen-1-ol (2-Ool). The females prefer the males producing a maximum amount of these sex pheromones (Stahr and Seidelmann, 2016). Therefore, by reducing the production of these sex pheromones, copulation and hence locust reproduction can be reduced. Recent molecular studies involving the blockade of pheromone signaling through receptors of the olfactory system could prove an effective approach for locust management. The pheromone, 4-vinylanisole (4VA) plays a role in the aggregation of *L. migratoria* and attracts the gregarious as well as solitary locusts strongly. Moreover, Knockout of OR35 (an olfactory receptor of 4VA) with CRISPR-Cas9 significantly impairs the electrophysiological responses of the antennae and 4VA attractiveness (Guo et al., 2020). Hence, with 4VA, the aggregated locusts can be collected and destroyed or by the inhibition of attractiveness toward this pheromone, the aggregation of locusts can be prevented. Heritable mutagenesis induced in *L. migratoria* using CRISPR/Cas9 system disrupts the gene expressing the odorant receptor co-receptor (*Orco*). Under crowding conditions, the *Orco* mutants lose an attraction response to aggregation pheromones (Li et al., 2016).

4.5.3.3 INSECT GROWTH REGULATORS (IGRS)

The IGRs may adversely affect growth and development of insects by regulating or inhibiting specific metabolic processes. These may be synthetic chemicals or natural compounds of plant origin. Many analogs or mimics of hormones indigenous to insects are now available and used against insects (Pener, 2012). The IGRs are categorized into (1) juvenile hormones (JHs) and their analogues (JHAs), (2) moulting hormone analogues (MHAs) and antimoulting hormone analogues, and (3) chitin synthesis inhibitors (CSIs). Commercially, JHAs and CSIs are the most exploited IGRs followed by MHAs while the work on other IGRs is still in its infancy (Singh et al., 2013). Some of the IGRs are also active against locusts and are used successfully in locust management programs. JH analogs may prolong the insect's immature stage and keep the insect in potentially injurious stage longer than usual. Therefore, any chemical capable to stop the production of JH could result in larval metamorphosis prematurity and production of sterile adults. This strategy could prove beneficial in locust outbreak management. An anti-JHA, fluvastatin, a synthetic inhibitor of 3-hydroxy-3-methylglutaryl-CoA reductase, suppresses the biosynthesis of JH by the corpora allata of locust under in vitro conditions but exhibits this impact in vivo only for 1 weak (Debernard et al., 1994). Ketokonazole is a synthetic imidazole derivative suppresses the biosynthesis of ecdysteroid in ovarian follicle cells of *L. migratoria.* This in turn impairs the locust reproduction and helps in the locust control. The anti-JHA, KK-42 (1-benzyl-5-[(E)-2,6-dimethyl-1,5-heptadienyl] imidazole), the phenyl derivatives of substituted imidazoles, inhibit juvenile hormone (JH) biosynthesis and delays or inhibit ecdysteroid synthesis in the desert locust females, *S. gregaria* (Wang and Schnal, 2001). The antibiotic, cycloheximide (Acti-dione) is an inhibitor of RNA, protein synthesis isolated initially from *Streptomyces griseus* and exhibits JH and anti-JH-like activity. It interferes with the hormonal regulation of development processes of *L. migratoria* and *S. gregaria* (Phillips and Loughton, 1979; Eid et al., 1982). Cycloheximide exerts toxic effects on *S. gregaria* fourth instar nymphs as well as adults. It affects the moulting and causes the death of the nymphs within a few days. Moreover, cycloheximide at lower doses induces solitary tendency in the nymphs (Tanani, 2018). The CSIs include the benzoylphenyl ureas, for example, diflubenzuron (Dimilin®, Chemtura AgroSolutions) and novaluron (Rimon®, Chemtura AgroSolutions). The benzoylphenyl ureas inhibit chitin synthesis and are effective for immature insect stages (Ishaaya and Casida, 1974; Ishaaya et al., 1998). The diflubenzuron or novaluron application to the second instars of *L. migratoria migratorioides* simultaneously with *B. bassiana* treatment exhibits an additive inhibitory effect and a cumulative inhibitory effect when the fungus is applied first and IGRs after 48 h (Bitsadze et al., 2013). Systemic RNA interference (RNAi), a reverse genetics approach is a more effective and an eco-friendly technique for locust pest control (Luo et al., 2013; Santos et al., 2014). The function of Methoprene-tolerant (Met), a JH receptor is in maturation of ovaries, vitellogenesis, and biosynthesis of ecdysteroid in females of *S. gregaria.* The RNA interference (RNAi) mediated *S. gregaria* Methoprene-tolerant (*SgMet*) knockdown decreases *S. gregaria* insulin-related peptide (*SgIRP*) and enhancement of *S. gregaria* neuroparsin (*SgNP*) 3 and 4 transcript levels in the fat body, delays copulation behavior display with virgin males and prevents adult *dsSgMet* injected female locusts to egg laying (Gijbels et al., 2019). Incorporation of IGRs, diflubenzuron, and teflubenzuron into bran baits could serve as a means of locust control. Both of these IGRs cause abortive moult and most survivors of *S. gregaria* developed twisted or misshapen wings (Wakgari, 1997). The benzoyl urea insect growth regulator, triflumuron is useful in barrier treatment against gregarious hopper bands of *L. migratoria capito* (Scherer and MA, 1993). Growth blocking peptides (GBPs) are the insect cytokines participating in the activation of immune system and retardation of insect growth. These peptides cause hemocyte spreading in vitro, the first step in the activation of hemocytes against infection in several species of insects. A locust growth blocking peptide (GBP) also known as locust hemocyte spreading peptide (locust GBP) in *L. migratoria* elicits the depletion of hemocytes, prolongs the larval growth phase and postpones the adult molting (Duressa et al., 2015).

4.6 INTEGRATED PEST MANAGEMENT (IPM)

The term IPM was used in 1967 by Smith and van den Bosch (Smith et al., 1967) and was accepted in 1969 by the US National Academy of Sciences (1969) (National Academy of Sciences Publication, 1969). During 1970s and 1980s, the governments throughout the globe adopted IPM as the main policy, research and extension strategy. It involves the appropriate utilization of all the available pest management techniques and methods to reduce the pest populations with minimum possible use of pesticides and other interventions in an economic and eco-friendly manner. IPM puts stress on crop growth, minimizing the damage and promoting natural pest management (FAO, 2012). Recent research lays emphasis on adopting an IPM approach as an ultimate solution to locust control. In the past few years, great development has been done and the principles of IPM have been implemented to control locusts. The latter involves: (1) a good understanding of population dynamics, (2) monitoring and early warning systems for better management of the evolution of populations and action in time, (3) models to assess the effectiveness and expenses of various control operations, (4) action thresholds, (5) a range of insecticides providing the opportunity to select them appropriately as per the requirement, (6) biological products, (7) follow-up methods to observe impact of locust control operations on the environment and (8) Additionally promoting and funding research, organizing symposia, workshops, etc. (Lecoq, 2010). Therefore, a complete range of complementary methods (viz. monitoring and early warning, physical, chemical and biological) are available in the IPM. Furthermore, the participation of the local people matters a lot to the implementation of pest control measures. Public outrage poses a great threat to IPM programs. Sometimes, it becomes almost impossible to gain their trust and support. In order to convince them, a well financially supported public information program should be launched at the launch of each pest control program. The strategies employed for pest management through IPM must ensure good public and environment health (Vreysen ct al., 2007).

4.7 REGIONAL, NATIONAL, AND INTERNATIONAL COOPERATION

The development of regional, national, and international cooperation is necessary for a more effective, sustainable, and long-term management of locust outbreak. This cooperation can be ethical, financial, educational as well as political eradicating the hinderance in the application of locust management strategies. Various regional, national, and international organizations act in coordination with one another, so the locust management programs keep progressing incessantly. Many insurance schemes are advertised and funds are released to mitigate the pest management risks (Huis, 2007). FAO plays major role in locust management by hosting the locust control committee involving various nations, helps the regional organizations coordination, sponsors the regional locust commissions, supports financial donors coordination meetings in the case of emergency campaigns, follows the locust situation and forecasts the situation (Pantenius and Butrous, 2017). One of the most significant facts is the launch of the huge Emergency Prevention System (EMPRES) program by the FAO to control the desert locust (Lecoq, 2003). The FAO launches various research programs and official sites to gather, save, and convey updated information regarding the locust situation (FAO, 2020). The participation of the national governments and agencies is also needed to make people aware of the locust situation to forecast the locust outbreak, to convey control strategies, to release funds, contingencies and to monitor the utilization of funds and implementation of locust control operations (Contingency

Plan for Desert Locust Invasions, Outbreaks and Upsurges Government of India, Ministry of Agriculture & Farmers Welfare 2019). The FAO along with national governments works for locust management. For example, the desert locust crisis Somalia action plan (January–December 2020) part of FAO's regional appeal for rapid response and anticipatory action in the greater horn of Africa. The co-sponsorship of FAO, United Nations Development Program (UNDP), United Nations Environmental Program (UNEP), and the World Bank (WB) establishes the global IPM facility for pest management (Maredia et al., 2003). Further, the establishment and enforcement of legislations by authorities bring desirable outcomes. These legislations are related to the ethics, edibility, farming, marketing, and outbreaks of insects including locusts (Baiano, 2020; Schiel et al., 2020; Abdel-Magid and Al-Zawahry, 2020). However, mere establishment and implementation of legislations is not enough for locust control (Story et al., 2005).

4.8 CONCLUSION

Currently, improvements in monitoring and early warning systems, physical, chemical, and biological control methods revolutionize the locust control. These improvements are because of the introduction and application of recent scientific techniques and tools. Improvements in communication modes, use of geospatial data (GIS and GPS), discovery and application of novel remote sensing satellites makes it possible to monitor and forecast the situation before and after the locust outbreaks so that appropriate measures can be taken in time. The conventional physical control measures accompanied with latest scientific instruments become more effective against locusts. Novel formulations and application techniques increase the effectiveness, and reduce the time and amount of application of synthetic chemicals to lager area. Biological locust control strategies, such as the release of predators, use of locusticidal microbes (fungi, bacteria, microsporidia, nematodes, and viruses) and biochemicals assist in overcoming the limitations of chemical control methods. Recent molecular techniques improve the antilocust action and help in understanding the mechanism of action of the biocontrol agents. In spite of various advantages, the application of biological control agents suffers from many limitations such as very slow mode of action and biosafety concerns. Moreover, only a few microsporidia, nematodes, viruses, and botanicals have been evaluated for their antilocust efficacy and they need more appraisals. Out of all the locust control strategies, IPM is the most effective and sustainable approach to combat locust pest outbreaks. It involves the participation of all the locust control strategies along with various legislations and involvement of national and international cooperation (scientific, political, financial as well as social). However, the locust control strategies mostly revolve around *S. gregaria*, *L. migratoria*, and *C. terminifera* and are scarce for other locusts, such as *D. maroccanus*, *S. piceifrons piceifrons* that need concern. Hopefully in future, the combined efforts of regional, national, and international organizations specifically of the FAO, the World Health Organization (WHO), the World Bank (WB), etc. will succeed in a more effective, eco-friendly and sustainable prevention and control of locust outbreaks.

KEYWORDS

- **locusts**
- **outbreaks**
- **management strategies**

REFERENCES

Abbassi, K.; Atay-Kadiri, Z.; Ghaout, S. Biological Effects of Alkaloids Extracted from Three Plants of Moroccan Arid Areas on the Desert Locust. *Physiol. Entomol.* **2003**, *28* (3), 232–236.

Abdalla, A. M. Studies on the Insecticidal Properties of Extracts from Roots of *Mucuna pruriens* (Fabaceae) Against Migratory Locust, *L. migratoria* and Desert Locust, *Schistocerca gregaria*, Doctoral Dissertation, University of Khartoum, **2004**.

Abdelatti, Z. A. S.; Hartbauer, M. Plant Oil Mixtures as a Novel Botanical Pesticide to Control Gregarious Locusts. *J. Pest. Sci.* **2020,** *93*, 341–353.

Abdellaoui, K.; Miladi, M.; Mkhinini, M.; Boughattas, I.; Hamouda, A. B.; Hajji-Hedfi, L.; Tlili, H.; Acheuk, F. The Aggregation Pheromone Phenylacetonitrile: Joint Action with the Entomopathogenic Fungus *Metarhizium anisopliae* Var. *acridum* and Physiological and Transcriptomic Effects on *Schistocerca gregaria* Nymphs. *Pestic. Biochem. Phys.* **2020,** 104594.

Abdel-Magid, I. M.; Al-Zawahry, A. E. Preconditions and Requirements for Successful Environmental Policies in the Sultanate of Oman, the Sudan and Egypt1. *Education* **2020,** *3431*, 90.

Ali, S.; Huang, Z.; Ren, S. X.; Bashir, M. H.; Afzal, M.; Long. T. Production and Extraction of Extracellular Lipase from Entomopathogenic Fungus *Metarhizium anisopliae* (Clavicipitaceae: Hypocreales). *Pak. J. Zool.* **2009,** *41* (5), 341–347.

Amiel, H.; Chohan, A. B.; Snibson, G. R.; Vajpayee, R. A Typical Fungal Sclerokeratitis. *Cornea* **2008,** *27* (3), 382–383.

Anonymous. Locust Alarm, Scientific Rigour and a Debate Over Risky Disease Experiments. *Nature* **2020,** *577*, 602–603.

Arroyo, A. S.; Martinez, G. J.; Pereda, S. M.; Sierra, F. D. First Breeding Record of Grasshopper Sparrow in Northern Durango, Mexico. Huitzil. *Rev. Mex. de Ornitol.* **2019,** *20* (2), 1–7.

Aw, K. M.; Hue, S. M. Mode of Infection of *Metarhizium* Spp. Fungus and Their Potential as Biological Control Agents. *J. Fungi.* **2017,** *3* (2), 30.

Awad, H. H.; Ghazawy, N. A.; Rahman, K. A. Impact of Farnesol on the Food Consumption and Utilization, Digestive Enzymes and Fat Body Proteins of the Desert Locust *Schistocerca gregaria* Forskål (Orthoptera: Acrididae). *Afr. Entomol.* **2013,** *21* (1), 126–131.

Bagari, M.; Bouhaimi, A.; Ghaout, S.; Chihrane, J. The Toxic Effects of *Nerium Oleander* on Larvae of the Desert Locust *Schistocerca gregaria* (Forskål, 1775) (Orthoptera, Acrididae). *Zool. Baetica.* **2013,** *24*,193–203.

Bagari, M.; Bouhaimi, A.; Ghaout, S.; Chihrane, J. Toxic Effects of *Nerium Oleander* on the Reproduction of the Desert Locust *Schistocerca gregaria* (Forskål 1775, Orthoptera, Acrididae). *Zool. baetica.* **2015,** *26*, 153–166.

Bai, N. S.; Remadevi, O. K.; Sasidharan, T. O.; Balachander, M.; Dharmarajan, P. Cuticle Degrading Enzyme Production by Some Isolates of the Entomopathogenic Fungus, *Metarhizium anisopliae* (Metsch.). J. *Bio-Sci.* **2012,** *20*, 25–32.

Baiano, A. Edible Insects: An Overview on Nutritional Characteristics, Safety, Farming, Production Technologies, Regulatory Framework, and Socio-Economic and Ethical Implications. *Trends Food Sci. Tech.* **2020,** *100*, 35–50.

Baker, B. P.; Benbrook, C. M.; Groth, E. III.; Benbrook, K. L. Pesticide Residues in Conventional, Integrated Pest Management (IPM)-Grown and Organic Foods: Insights from Three US Data Sets. *Food Addit. Contam.* **2002,** *19* (5), 427–446.

Baker, G. L.; Capinera, J. L. Nematodes and Nematomorphs as Control Agents of Grasshoppers and Locusts. *Mem. Entomol. Soc. Can.* **1997,** *129* (S171), 157–211.

Banik, A.; Mondal, M. F.; Khan, M. M.; Ahmed, S. R.; Hasan, M. M. Screening and Potent Applicability Analysis of Commonly Used Pesticides Against Desert Locust: An Integrative Entomo-Informatics Approach. *BioRxiv.* **2020,** 1–29.

Bashir, M.; Hassanali, A.; Korena, H.; Bashir, A. W. Semi-Field Evaluation of the Effects of Sub-Lethal Doses of Pesticides, with and without Adult *Schistocerca gregaria* Pheromone, on Hoppers. *Entomol. Exp. Appl.* **2016,** *160* (1), 72–76.

Belayneh, Y. T. Acridid Pest Management in the Developing World: A Challenge to the Rural Population, a Dilemma to the International Community. *J. Orthoptera Res.* **2005,** *14* (2), 187–195.

Bensimon, A.; Zinger, S.; Gerassi, E.; Hauschner, A.; Harpaz, I.; Sela, I. "Dark Cheeks," a Lethal Disease of Locusts Provoked by a *Lepidopterous baculovirus*. *J. Invertebrr Pathol.* **1987,** *50* (3), 254–260.

Bitsadze, N.; Jaronski, S.; Khasdan, V.; Abashidze, E.; Abashidze, M.; Latchininsky, A.; Samadashvili, D.; Sokhadze, I.; Rippa, M.; Ishaaya, I.; Horowitz, A. R. Joint Action of *Beauveria Bassiana* and the Insect Growth Regulators Diflubenzuron and Novaluron, on the Migratory Locust, *Locusta migratoria*. *J. Pest Sci.* **2013,** *86* (2), 293–300.

Bonnal, V.; lagnaoui, S.; Dolley, T.; Tonneau, J.-P.; Lecoq, M.; Vassal, J.-M. A Monitoring System for Preventive Control of Desert Locust in West Africa. Scientific and Technical Information and Rural Development IAALD XIIIth World Congress, Montpellier, 2010.

Brunner-Mendoza, C.; Navarro-Barranco, H.; León-Mancilla, B.; Pérez-Torres, A.; Toriello, C. Biosafety of an Entomopathogenic Fungus *Isaria fumosorosea* in an Acute Dermal Test in Rabbits. *Cutan. Ocul. Toxicol.* **2017,** *36* (1), 12–18.

Brunner-Mendoza, C.; Reyes-Montes, M. D.; Moonjely, S.; Bidochka, M. J.; Toriello, C. A Review on the Genus *Metarhizium* as an Entomopathogenic Microbial Biocontrol Agent with Emphasis on Its Use and Utility in Mexico. *Biocontrol Sci. Technol.* **2019,** *29* (1), 83–102.

Bryceson, K. P. Discrimination of Small Areas of Vegetation Growth After Localized Rain in Semi-Arid Environments Using Landsat. In *Proceed. Third Australian Remote Sens. Conf.*, Queensland. **1984;** pp 541–550.

Bryceson, K. P. The Use of Landsat MSS Data to Determine the Distribution of Locust Egg Beds in the Riverina Region of New South Wales, Australia. *Int. J. Remote Sens.* **1989,** *10* (11), 1749–1762.

Bryceson, K. P.; Wright, D. E. an Analysis of the 1984 Locust Plague in Australia Using Multitemporal Landsat Multispectral Data and a Simulation Model of Locust Development. *Agric. Ecosyst. Environ.* **1986,** *16* (2), 87–102.

Bukhari, T.; Middelman, A.; Koenraadt, C. J.; Takken, W.; Knols, B. G. Factors Affecting Fungus-Induced Larval Mortality in *Anopheles Gambiae* and *Anopheles Stephensi*. *Malar. J.* **2010,** *9* (1), 1–15.

Burges, H. D.; Hussey, N. W. *Microbial Control of Insects and Mites*; Academic Press, **1971;** p 86.

Burgner, D.; Eagles, G.; Burgess, M.; Procopis, P.; Rogers, M.; Muir, D.; Priest, M. Disseminated Invasive Infection Due to *Metarhizium anisopliae* in an Immunocompromised Child. *J. Clin. Microbiol.* **1998,** *36* (4), 1146–1150.

Calvão, T.; Pessoa, M. F. Remote Sensing in Food Production-A Review. Emir. *J. Food Agric.* **2015,** 27 (2 (SI)), 138–151.

Cease, A. J.; Elser, J. J.; Ford, C. F.; Hao, S.; Kang, L.; Harrison, J. F. Heavy Livestock Grazing Promotes Locust Outbreaks by Lowering Plant Nitrogen Content. *Science* **2012,** 335 (6067), 467–469.

Cherlet, M. RAMSES, Reconnaissance and Management System of the Environment of *Schistocerca*. Coordinated Development Programme Contributing to Desert Locust Preventive Control Operations by Improving Monitoring and Forecasting, Through Optimization of Remote Sensing and Integration in a GIS into a User Dedicated Tool. Food and Agriculture Organization of the United Nations, Rome, 1993.

Cherlet, M.; Di Gregorio, A. Calibration and Integrated Modelling of Remote Sensing Data for Desert Locust Habitat Monitoring. Project ECLO/INT/004/BEL. Project GCP/INT/439/BEL. Food and Agriculture Organization of the United Nations, Rome, 1991.

Chester, B.; Stotzky, G. Temperature-Dependent Cultural and Biochemical Characteristics of Rhamnose-Positive *Yersinia enterocolitica. J. Clin. Microbiol.* **1976,** *3* (2), 119–127.

Chi, T. An Analysis of the Historical Records of Locust Outbreaks in China. *Chinese J. Agr. Res.* **1950,** *1*, 57–66.

Chilali, F.; Benrima, A. Evaluation of the Insecticide Activity of Schinus Molle on the Migratory Locust *Locusta migratoria* (Acrididae, Oedipodinae). *AgroBiologia* **2018,** *8* (1), 879–885.

Chouvenc, T.; Su, N.-Y.; Kenneth Grace, J. Fifty Years of Attempted Biological Control of Termites-Analysis of a Failure. *Biol. Control.* **2011,** *59* (2), 69–82.

Comisión Federal Para La Protección Contra Riesgos Sanitarios (COFEPRIS). X Reglamento En Material De Registros, Autorizaciones De Importación Y Exportación Y Certificados De Exportación De Plaguicidas, Nutrientes Vegetales Y Sustancias Y Materiales Tóxicos O Peligrosos [Regulations on Registration Material, Import and Export Authorizations and Export Certificates for Pesticides, Plant Nutrients and Toxic Or Dangerous Substances and Materials]. Diario Oficial De La Federación, Mexico, 2005.

Contingency Plan for Desert Locust Invasions, Outbreaks and Upsurges Government of India, Ministry of Agriculture & Farmers Welfare, (Department of Agriculture, Cooperation and Farmers Welfare), Directorate of Plant Protection, Quarantine & Storage, NH-IV, Faridabad-121001, 2019.

COPR (Centre for Overseas Pest Research). The Locust and Grasshopper Agricultural Manual, Centre for Overseas Pest Research, London, 1982.

Cortés-Martínez, C. I.; Chavarría-Hernández, N. Production of Entomopathogenic Nematodes in Submerged Monoxenic Culture: A Review. *Biotechnol. Bioeng.*, 2020.

Courcoux, G. Invasions of Locusts: A Lasting Impact. Scientific News of the Institut De Recherche Pour Le Développement, 2012.

Cressman, K. *Desert Locust. Biological and Environmental Hazards, Risks and Disasters*, 1st ed.; **2016;** pp 87–105.

Cressman, K. SWARM: A Geographic Information System for Desert Locust Forecasting. In *New Strategies in Locust Control*; Krall, S., Peveling, R., Badiallo, D., Eds.; Birkhäuser, Basel, Switzerland, **1997;** pp 27–36.

Cullen, D. A.; Cease, A. J.; Latchininsky, A. V.; Ayali, A.; Berry, K.; Buhl, J.; De Keyser, R.; Foquet, B.; Hadrich, J. C.; Matheson, T.; Ott, S. R. From Molecules to Management: Mechanisms and Consequences of Locust Phase Polyphenism. *Adv. Insect Physiol.* **2017,** *53*, 167–285.

Daniel, G. J.; Haig, K. Laser Inactivation of Fasciclin-I Disrupts Axon Adhesion of Grasshopper Pioneer Neurons. *Nature* 1990, *348* (6301), 548–550.

de Faria, M. R., Wraight, S. P. Mycoinsecticides and Mycoacaricides: A Comprehensive List with Worldwide Coverage and International Classification of Formulation Types. *Biol. Control.* **2007,** *43* (3), 237–256.

De García, M. C.; Arboleda, M. L.; Barraquer, F.; Grose, E. Fungal Keratitis Caused by *Metarhizium anisopliae* Var. *anisopliae. J. Med. Vet. Mycol.* **1997,** *35* (5), 361–363.

De Vreyer, P.; Guilbert, N.; Mesple-Somps, S. The 1987–89 Locust Plague in Mali: Evidences of the Heterogeneous Impact of Income Shocks on Education Outcomes. No DT/2012/05, Working Papers, DIAL (Développement, Institutions et Mondialisation), 2012.

Debernard, S.; Rossignol, E; Couillaud, E. The HMG-CoA Reductase Inhibitor Fluvastatin Inhibits Insect Juvenile Hormone Biosynthesis. *Gen. Comp. Endocrinol.* **1994,** *95* (1), 92–98.

Deveson, E. D. The Search for a Solution to Australian Locust Outbreaks: How Developments in Ecology and Government Responses Influenced Scientific Research. *Hist. Rec. Aust. Sci.* **2011,** *22* (1), 1–31.

Diedhiou, P. M.; Badji, K.; Faye, R.; Samb, P. I. Stability of *Metarhizium acridum* Based Biopesticide in Operational Conditions in Senegal. *Curr. Agric. Res. J.* **2014,** *2* (2), 83–88.

Dobson, H. Advances in Locust Spraying Technology. *Insect Sci. Appl.* **1999,** *19* (4), 355–368.

Duressa, T. F.; Boonen, K.; Hayakawa, Y.; Huybrechts, R. Identification and Functional Characterization of a Novel Locust Peptide Belonging to the Family of Insect Growth Blocking Peptides. *Peptides* **2015,** *74*, 23–32.

Eguchi, H.; Toibana, T.; Hotta, F.; Miyamoto, T.; Mitamura, Y.; Yaguchi, T. Severe Fungal Sclerokeratitis Caused by *Metarhizium anisopliae*: A Case Report and Literature Review. *Mycoses* **2015,** *58* (2), 88–92.

Eid, M. A.; Salem, M. S.; El-Ibrashy, M. T.; Abdel-Hamid, M. Effects of Precocene II, Cycloheximide and C16-JH on the Growth Rate and Adult Morphometrics of *Schistocerca gregaria* Forsk. *Bull. Entomol. Soc. Egypt (Econ. Ser.)* 1982, *13*, 95–105.

El-Ibrashy, M. T.; Abdel-Hamid, M.; El-Refai, A. L. Ecdysones and Plant Growth Regulators Induce Solitarious Characters and Reduce Fertility in the Desert Locust. *Entomol. Exp. Appl.* **1976,** *19* (3), 214–220.

Erlandson, M. A.; Streett, D. A. Entomopoxviruses Associated with Grasshoppers and Locusts: Biochemical Characterization. *Mem. Entomol. Soc Can.* **1997,** *129* (S171), 131–146.

Fang, W.; Lu, H. L.; King, G. F.; Leger, R. J. Construction of a Hypervirulent and Specific Mycoinsecticide for Locust Control. *Sci. Rep.* **2014,** *4*, 7345.

FAO 2020. FAO Locust Watch, Food and Agriculture Organization of the United Nations, Rome, Italy. http://www.fao.org/ag/locusts.

FAO. Evaluation of Field Trials Data on the Efficacy and Selectivity of Insecticides on Locusts and Grasshoppers. Report to FAO by the Pesticide Referee Group. Eighth Meeting. Rome, 11–14 October 1999. Food and Agriculture Organization of the United Nations, Rome, 1999.

FAO. FAO Desert Locust Control Committee, Thirty-Ninth Session. Rome: Food and Agriculture Organization of the United Nations. 2009.

FAO. Integrated Pest Management. http://www.fao.org/agriculture/crops/core-themes/theme/pests/ipm/en/. 2012.

FAO. Promoting the Contribution of Edible Forest Insects in Assuring Food Security. FAO Forestry Department Programme. Forest Economy, Policy and Product Divisions. Rome: FAO, 2010.

Franc A. Impact des Transformations Mésologiques Sur La Dynamique Des Populations Et La Grégarisation Du Criquet Nomade Dans Le Bassin De La Sofia (Madagascar) (Doctoral Dissertation, Montpellier 3), 2007.

Fu, X. J.; Hunter, D. M.; Shi, W. P. Effect of *Paranosema* (*Nosema*) *locustae* (Microsporidia) on Morphological Phase Transformation of *Locusta migratoria manilensis* (Orthoptera: Acrididae). *Biocontrol Sci. Technol.* **2010,** *20* (7), 683–693.

Gabarty, A.; Salem, H. M.; Fouda, M. A.; Abas, A. A.; Ibrahim, A. A. Pathogenicity Induced by the Entomopathogenic Fungi *Beauveria Bassiana* and *Metarhizium Anisopliae* in *Agrotis Ipsilon* (Hufn.). *J. Radiat. Res. Appl. Sci.* 2014, *7* (1), 95–100.

García de León, S.; Mier, T. Visión General De La Producción Y Aplicación De Bioplaguicidas En México. Sociedades Rurales, Producción Y Medio Ambiente. **2010,** *10* (20), 37–63.

Ghoneim, K.; Hamadah, K.; El-Hela, A.; Mohammad, A. H.; Amer, M. Efficacy of *Nigella sativa* (Ranunculaceae) Extracts on Adult Performance and Phase Transition of the Desert Locust *Schistocerca gregaria* (Orthoptera: Acrididae). *Int. J. Res. Stud. Zool.* **2015,** *1* (2), 29–44.

Ghoneim, K. E.; Hamadah, K. S.; El-Hela, A. A. Acetylcholinesterase Activity in the Desert Locust *Schistocerca gregaria* (Acrididae) (Forsk.) as a Response to the Action of the Wild Herb *Fagonia bruguieri* DC. (Zygophyllaceae) Extracts. J. *Entomol. Res. Soc.* **2012,** *14* (2), 87–97.

Gijbels, M.; Lenaerts, C.; Broeck, J. V.; Marchal, E. Juvenile Hormone Receptor Met Is Essential for Ovarian Maturation in the Desert Locust, *Schistocerca gregaria. Sci. Rep.* **2019,** *9* (1), 1–12.

Glare, T.; Caradus, J.; Gelernter, W.; Jackson, T.; Keyhani, N.; Köhl, J.; Marrone, P.; Morin, L.; Stewart, A. Have Biopesticides Come of Age? *Trends Biotechnol.* **2012,** *30* (5), 250–258.

Goettel, M. S.; Johnson, D. L. Microbial Control of Grasshoppers and Locusts. Eds. *Mem. Entomol. Soc. Can.* **1997.** 171.

Goodman, A. L.; Lockhart, S. R.; Lysen, C. B.; Westblade, L. F.; Burnham, C. A. D.; Burd, E. M. Two Cases of Fungal Keratitis Caused by *Metarhizium anisopliae. Med. Mycol. Case Rep.* **2018,** *21*, 8–11.

Goodman, B. Research Community Swats Grasshopper Control Trial. *Science* **1993,** *260* (5110), 887–887.

Guerrero-Guerra, C.; del Roció Reyes-Montes, M.; Toriello, C.; Hernández-Velázquez, V.; Santiago-López, I.; Mora-Palomino, L.; Calderón-Segura, M. E.; Fernández, S. D.; Calderón-Ezquerro, C. Study of the Persistence and Viability of *Metarhizium acridum* in Mexico's Agricultural Area. *Aerobiologia* **2013,** *29* (2), 249–261.

Guo, X.; Yu, Q.; Chen, D.; Wei, J.; Yang, P.; Yu, J.; Wang, X.; Kang, L. 4-Vinylanisole Is an Aggregation Pheromone in Locusts. *Nature* 2020, *584* (7822), 584–588.

Hansson, B. S.; Stensmyr, M. C. Evolution of Insect Olfaction. *Neuron* **2011,** *72* (5), 698–711.

Hassanali, A.; Bashir, M. O. Insights for the Management of Different Locust Species from New Findings on the Chemical Ecology of the Desert Locust. *Insect Sci. Appl.* **1999,** *19* (4), 369–376.

Hassanali, A.; Njagi, P. G. N.; Bashir, M. O. Chemical Ecology of Locusts and Related Acridids. *Annu. Rev. Entomol.* **2005,** *50*, 223–245.

Hastings, J.; Latchininsky, A. V.; Schell, S. P. Sustainability of Grasshopper Management and Support Through CARMA. In *Proceed. 42nd Annu Hawai'i Int. Conf. Syst. Sci., Los Alamitos*, CA: *IEEE Comp. Soc*. 2009; pp 1–10.

Hernandez, C. E.; Guerrero, I. E.; Hernandez, G. A.; Solis, E. S.; Guzman, J. C. Catalase Overexpression Reduces the Germination Time and Increases the Pathogenicity of the Fungus *Metarhizium anisopliae*. *Appl. Microbiol. Biotechnol*. **2010,** *87* (3), 1033–1044.

Hielkema, J. U. Application of Landsat Data in Desert Locust Survey and Control. AGP: LCC/77/11, FAO, Rome, Italy. 1977.

Hielkema, J. U. Remote Sensing Techniques and Methodologies for Monitoring Ecological Conditions for Desert Locust Population Development. FAO/USA ID Final Report. FAO, Rome, Italy. 1980.

Hough, P. Sustainable Food Production Includes Human and Environmental Health (Vol. 3), **2014;** pp 1–41.

http://carma.unk.edu/

http://www.fao.org/ag/locusts/en/activ/DLIS/satel/index.html.

Hu, X.; Xiao, G.; Zheng, P.; Shang, Y.; Su, Y.; Zhang, X.; Liu, X.; Zhan, S.; Leger, R. J.; Wang, C. Trajectory and Genomic Determinants of Fungal-Pathogen Speciation and Host Adaptation. *Proc. Natl. Acad. Sci*. **2014,** *111* (47), 16796–16801.

Hunter, D. M. Mycopesticides as Part of Integrated Pest Management of Locusts and Grasshoppers. *J. Orthoptera Res*. **2005,** *14* (2), 197–201.

Hurst, M. H.; Glare, T. R. Novel Bacteria and Uses Thereof. Patent No. PCT/NZ2007/000146. Australia, 2006.

Ingrisch, S.; Rentz, D. C. F. Orthoptera: Grasshoppers, Locusts, Katydids, Crickets. In *Encyclopedia of Insects*; Academic Press, **2009;** pp 732–774.

Ishaaya, I.; Casida, J. E. Dietary TH 6040 Alters Cuticle Composition and Enzyme Activity of House Fly Larval Cuticle. *Pesticide Biochem. Physiol*. **1974,** *4* (4), 484–490

Ishaaya, I.; Horowitz, A. R. Insecticides with Novel Modes of Action: An Overview. In: Ishaaya I, Degheele D (eds) Insecticides with Novel Modes of Action, Mechanism and Application. Springer, Berlin, **1998;** pp 1–24.

Isman, M. B. Botanical Insecticides in the Twenty-First Century-Fulfilling Their Promise?. *Annu. Rev. Entomol*. **2020,** *65*, 233–249.

Jani, B. R.; Rinaldi, M. G.; Reinhart, W. J. an Unusual Case of Fungal Keratitis: *Metarrhizium anisopliae*. *Cornea* 2001, *20* (7), 765–768.

Jaoko, V.; Nji Tizi Taning, C.; Backx, S.; Mulatya, J.; Van den Abeele, J.; Magomere, T.; Olubayo, F.; Mangelinckx, S., Werbrouck, S. P.; Smagghe, G. The Phytochemical Composition of *Melia volkensii* and Its Potential for Insect Pest Management. *Plants* **2020,** *9* (2), 143.

Jayaprakash, A.; Muralidharan, S.; Joseph, P.; Gokul Devendran. Cram of Crops and Control Measures Against Locust Swarm Attack. *Int. J. Curr. Res. Multidiscipl*. **2016,** *5* (9), 01–04.

Ji, R.; Xie, B. Y.; Li, D. M.; Li, Z.; Zhang, X. Use of MODIS Data to Monitor the Oriental Migratory Locust Plague. *Agric. Ecosys. Environ*. **2004,** *104* (3), 615–620.

Ji, R.; Xie, B. Y.; Li, Z.; Li, D. M.; Zhang, X.; Liu, T. J. Reasons and Characteristics of the Oriental Migratory Locust Plague Outbreak in Nandagang State Farm, Hebei Province. *Entomol. Knowl*. **2002,** 39, 430–433.

Jiang, W.; Peng, Y.; Ye, J.; Wen, Y.; Liu, G.; Xie, J. Effects of the Entomopathogenic Fungus *Metarhizium anisopliae* on the Mortality and Immune Response of *Locusta migratoria*. *Insects* **2020,** *11* (1), 36.

Kabaru, J. M.; Mwangi, R. W. Effect of Post-Treatment Temperature on the Insecticidal Activity of Neem, *Azadirachta indica* A. Juss. Seed Extract on *Schistocerca gregaria* (Forskal): A Preliminary Report. *Int. J. Trop. Insect. Sci.* **2000,** *20* (1), 77–79.

Karen, F.; Lee, B. A.; Paul, S.; William, A. B.; Michael, J. H. Cholinesterase Response in Native Birds Exposed to Fenitrothion During Locust Control Operations in Eastern Australia. *Environ. Toxicol. Chem.* **2006,** *25* (11), 2964–70.

Keeling, P. Five Questions About Microsporidia. *PLoS Pathog*. 2009, *5*, e1000489.

Kemassi, A.; Boual, Z.; Hadjseyd, A.; Bouziane, N.; Herouini, A.; Mensouri, K.; Bouras, N.; Ould, E. H.; Ould-El-Hadj, M. D. Biotoxic Effects of *Cleome arabica* L. (Capparidaceae) Extract from the Desert Locust *Schistocerca gregaria* (Forskål, 1775) (Orthoptera: Acrididae). *A. J. A. E.* **2018,** *8* (2), 79–98.

Khan, S. H. Recent Advances in Role of Insects as Alternative Protein Source in Poultry Nutrition. *J. Appl. Animal Res*. **2018,** *46* (1), 1144–1157.

Khusro, M.; Andrew, N. R.; Nicholas, A. Insects as Poultry Feed: A Scoping Study for Poultry Production Systems in Australia. *Worlds Poult. Sci. J.* 2012, *68* (3), 435–446.

Kim, K. H.; Kabir, E.; Jahan, S. A. Exposure to Pesticides and the Associated Human Health Effects. *Sci. Total Environ*. **2017,** *575*, 525–535.

Kooyman, C.; Bateman, R. P.; Langewald, J.; Lomer, C. J.; Ouambama, Z.; Thomas, M. Operational-Scale Application of Entomopathogenic Fungi for Control of Sahelian Grasshoppers. *Proceed. R. Soc. London B*. 1997, *264* (1381), 541–546.

Kritsky, G. The Insects and Other Arthropods of the Bible, the New Revised Version. *Am. Entomol.* **1997,** *44* (3), 183–188.

Lacey, L. A.; Grzywacz, D.; Shapiro-Ilan, D. I.; Frutos, R.; Brownbridge, M.; Goettel, M. S. Insect Pathogens as Biological Control Agents: Back to the Future. *J Invertebr pathol.* **2015,** *132*, 1–41.

Latchininsky A. V. Locusts. In: *Encyclopedia of Animal Behavior*, Breed, M. D., Moore, J., Eds., Vol. 2; Academic Press: Oxford, **2010;** pp 288–297.

Latchininsky, A. V. Moroccan Locust *Dociostaurus Maroccanus* (Thunberg, 1815): A Faunisticrarity Or an Important Economic Pest? *J. Insect Conservation* 1998, *2* (3–4), 167–178.

Latchininsky, A. V. Locusts and Remote Sensing: A Review. *J. Appl. Remote Sens*. **2013,** *7* (1), 075099.

Latchininsky, A. V.; Sergeev, M. G.; Childebaev, M. K.; Chernyakhovsky, M. E.; Lockwood, J. A.; Kambulin, V. E.; Gapparov, F. A. Acridids of Kazakhstan, Central Asia and Adjacent Territories. Association for Applied Acridology International, 2002.

Latchininsky, A. V.; Sivanpillai, R. Locust Habitat Monitoring and Risk Assessment Using Remote Sensing and GIS Technologies. Chapter 7 in *Integrated Management of Arthropod Pests and Insect Borne Diseases, Integrated Management of Plant Pests and Diseases*; Ciancio, A., Mukerji, K. G., Eds., Vol. 5, Springer Science+Business Media B. V.: Dordrecht, Heidelberg, London, New York, **2010;** pp 163–188.

Le Gall, M.; Overson, R.; Cease, A. J. A Global Review on Locusts (Orthoptera: Acrididae) and Their Interactions with Livestock Grazing Practices. *Front. Ecol. Evol.* **2019,** *7*, 263.

Le Gall, M.; Word, M. L. Thompson, N.; Beye, A.; Cease, A. J. Nitrogen Fertilizer Decreases Survival and Reproduction of Female Locusts by Increasing Plant Protein to Carbohydrate Ratio. *J. Anim. Ecol.* **2020,** *89* (10), 2214–2221.

Lecoq, M. Integrated Pest Management for Locusts and Grasshoppers: Are Alternatives to Chemical Pesticides Credible? *J. Orthoptera Res.* **2010,** *19* (1), 131–132.

Lecoq, M. Madagascar: Bilan De La Lutte Contre L'invasion Du Criquet Migrateur Et Perspectives Offertes Par Le Nouveau Centre National Antiacridien. FAO/AGP-P. Organisation Des Nations Unies Pour L'alimentation Et L'agriculture, Rome, 2000.

Lecoq, M. Papers Presented in a Symposium on "Desert Locust Control"* At the Eighth Arab Congress of Plant Protection Held in El-Beida, Libya, October 12-16, 2003 Desert Locust Threat to Agricultural Development and Food Security and FAO/International Role in Its Control.

Lednev, G.; Levchenko, M.; Kazartsev, I. Entomopathogenic Microorganisms in Locusts and Grasshoppers Populations and Prospects for Their Use for Control of This Pest Group. In *BIO Web of Conf.* EDP Sciences, **2020,** *21*, 00025.

Lednev, G. R.; Kryukov, V. Y.; Khodyrev, V. P.; Levchenko, M. A.; Duisembekov, B. A.; Sagitov, A. O.; Glupov, V. V. Dynamics of Mortality of the Migratory Locust Under Synchronous Infection with Entomopathogenic Fungi (*Beauveria bassiana*, *Metarhizium anisopliae*) and Bacteria *Pseudomonas* sp. *Contemp. Probl. Ecol.* 2008, *1* (2), 210–213.

Lednev, G. R.; Levchenko, M. V.; Uspanov, A. M.; Sagitov, A. O.; Pavlyushin, V. A. Modern Technologies and Plant Protection Products-A Platform for Innovative Development in the Agricultural Sector of Russia. **2018,** 95–97.

Li, Y.; Zhang, J.; Chen, D.; Yang, P.; Jiang, F.; Wang, X.; Kang, L. CRISPR/Cas9 in Locusts: Successful Establishment of an Olfactory Deficiency Line by Targeting the Mutagenesis of an Odorant Receptor Co-Receptor (Orco). *Insect Biochem. Mol. Biol.* **2016,** *79*, 27–35.

Liu, Z.; Shi, X.; Warner, E.; Ge, Y.; Yu, D.; Ni, S.; Wang, H. Relationship Between Oriental Migratory Locust Plague and Soil Moisture Extracted from MODIS Data. *Int. J. Appl. Earth Obs. Geoinf.* **2008,** *10* (1), 84–91.

Lockwood, J. A. Environmental-Issues Involved in Biological-Control of Rangeland Grasshoppers (Orthoptera. Acrididae) with Exotic Agents. *Environ. Entomol.* **1993,** *22* (3), 503–518.

Logue, C. M.; Sheridan, J. J.; McDowell, D. A.; Blair, I. S.; Hegarty, T. The Effect of Temperature and Selective Agents on the Growth of *Yersinia enterocolitica* Serotype O: 3 in Pure Culture. *J. Appl. Microbiol.* **2000,** *88* (6), 1001–1008.

Lomer, C. J.; Bateman, R. P.; Johnson, D. L.; Langewald, J.; Thomas, M. Biological Control of Locusts and Grasshoppers. *Annu. Rev. Entomol.* **2001,** *46* (1), 667–701.

Long, Z.; Hunter, D. M. Laboratory and Field Trials of Green Guard®; (*Metarhizium anisopliae* Var. *acridum*) (Deuteromycotina: Hyphomycetes) Against the Oriental Migratory Locust (*Locusta migratoria manilensis*) (Orthoptera: Acrididae) in China. *J. Orthoptera Res.* **2005,** *14* (1), 27–30.

Luo, Y.; Wang, X.; Wang, X.; Yu, D.; Chen, B.; Kang, L. Differential Responses of Migratory Locusts to Systemic RNA Interference Via Double-Stranded RNA Injection and Feeding. *Insect Mol. Biol.* **2013,** *22* (5), 574–583.

Ma, J. W.; Han, X. Z., Hasibagan,, Wang, C. L., Zhang, Y. J., Tang, J. Y., Xie, Z. Y. and Deveson, T. Monitoring East Asian Migratory Locust Plagues Using Remote Sensing Data and Field Investigations. *Int. J. Remote Sens.* **2005,** *26* (3), 629–634.

Magor, J. Desert Locust Forecasting GIS, a Current Technological Challenge. In Proc., 9th Entomological Congress of the Entomological Society of Southern Africa. Johannesburg, 1993.

Magor, J.; Pender, J. Desert Locust Forecasters'gis: A Researchers' View. In *New Strategies in Locust Control*; Krall, S., Peveling, R., Ba Diallo, D., Eds.; Birkhäuser: Basel, Switzerland, **1997;** pp 21–26.

Maleki, M.; Khorram, M. S. Bayesian Network and Pest Management: A Case Study of Trap Plants on Locust Population. *Adv. Environ. Biol.* **2010,** *4* (2), 147–151.

Mansour, S. A.; El-Sharkawy, A. Z.; Abdel-Hamid, N. A. Toxicity of Essential Plant Oils, in Comparison with Conventional Insecticides, Against the Desert Locust, *Schistocerca gregaria* (Forskål). *Ind. Crops Prod.* **2015,** *63*, 92–99.

Maredia, K. M.; Dakouo, D.; Mota-Sanchez, D. Editors. Integrated Pest Management in the Global Arena. CABI, 2003.

Mashtoly, T. A.; El-Zemaity, M. S.; Abolmaaty, A.; Abdelatef, G. M.; Marzouk, A. A.; Reda, M. Phylogenetic Characteristics of Novel *Bacillus weihenstephanensis* and *Pseudomonas* Sp. to Desert Locust, *Schistocerca gregaria* Forskål (Orthoptera: Acrididae). *Egypt J. Biol. Pest Control.* **2019,** *29* (1), 85.

Mcculloch, C.; Hunter, D. M. Identification and Monitoring of Australian Plague Locust Habitats from Landsat. *Remote Sens. Environ.* **1983,** *13* (2), 95–102.

Mcneill, M. R.; Hurst, M. R. H. *Yersinia* Sp. (Mh96)-A Potential Biopesticide of Migratory Locust *Locusta migratoria* L. *N. Z. Plant Prot.* **2008,** *61*, 236–242.

Meynard, C. N.; Lecoq, M.; Chapuis, M. P.; Piou, C. on the Relative Role of Climate Change and Management in the Current Desert Locust Outbreak in East Africa. *Glob. Chang Biol.* **2020,** *26* (7), 3753–3755.

Michelsen, A. Frequency Discrimination in Locust Ear by Means of 4 Groups of Receptor Cells. *Nature* **1968,** *220* (5167), 585.

Mier, T.; Olivares-Redonda, G.; Navarro-Barranco, H.; Pérez-Mejía, A.; Lorenzana, M.; Pérez- Torres, A.; Toriello, C. Acute Oral Intragastric Pathogenicity and Toxicity in Mice of *Paecilomyces fumosoroseus* Isolated from Whiteflies. *Anton. Leeuw. Int. J. G. Mol. Microbiol.* **2005,** *88* (2), 103–111.

Moharana, S.; Khuntia, P.; Das, S. S.; Das, R.; Panigrahi, S. A Systematic Review of Behavioral Aspects & Management Techniques to Control Locusts. *I. J. S. A. R. T.* **2020,** *6* (7), 2395–1052.

Motley, W. W.; Melson, A. T.; Mortensen, J. E. Pediatric *Metarrhizium anisopliae* Keratitis. *J. A. A. P. O. S.* **2011,** *15* (1), 101–103.

Mullié, W. C.; Guèye, Y. Does Bird Predation Enhance the Impact of Green Muscle® (*Metarhizium acridum*) Used for Grasshopper Control? *J. Orthoptera Res.* **2010,** *19*, 139–155.

Navratil, P. Object Based Locust Habitat Mapping Using Very High Resolution Satellite Data in the Southern Aral Sea Basin, Uzbekistan. Unpublished Master's Thesis, University of Regensburg, Regensburg, Germany, 2007.

Niassy, S.; Diarra, K.; Ndiaye, S.; Niassy, A. Pathogenicity of Local *Metarhizium anisopliae* Var. *acridum* Strains on *Locusta migratoria* Migratorioides Reiche and Farmaire and *Zonocerus variegatus* Linnaeus in Senegal. *Afr. J. Biotechnol.* **2011,** *10* (1), 28–33.

Nyam, V.-T.; Bong, C.-F. J.; King, J.-H. P. Control of Subterranean Termite Coptotermes Curvignathus (Isoptera: Rhinotermitidae) by Entomopathogen *Metarhizium anisopliae* Var. *anisopliae* Cultured in Liquid State Fermentation. *Am. J. Agric. Biol. Sci.* **2015,** *10* (1), 35.

Ocampo, V. R.; Caoili, B. L. Infection Process of Entomopathogenic Fungi *Metarhizium Anisopliae in* the *Tetranychus Kanzawai* (Kishida) (Tetranynichidae: Acarina). *Agrivita* **2013,** 35, 64.

Ostfeld, R. S.; Keesing, F. Ecology: Enhanced: Oh the Locusts Sang, Then They Dropped Dead. *Science* **2004,** *306* (5701), 1488–1489.

Ottesen P. S.; Butrous, M.; Corbett, M.; Fossland, S.; Jaffar, M.; Johannessen, B.; Sander, T. Field Tests on an Integrated Differential GPS Navigation and Spray Monitoring System for Aerial Desert Locust Control Operations. Technical Series-Desert Locust Field Research Stations (FAO), 1999.

Oulebsir-Mohandkaci, H.; Khemili-Talbi, S.; Benzina, F.; Halouane, F. Isolation and Identification of Entomopathogenic Bacteria from Algerian Desert Soil and Their Effects Against the Migratory Locust, *Locusta migratoria* (Linnaeus, 1758) (Orthoptera: Acrididae). *Egypt. J. Biol. Pest Control* **2015,** *25* (3), 739.

Pantenius, C.; Butrous, M. The FAO Commission for Controlling the Desert Locust in the Central Region (CRC). A Celebration of 50 Years of Service. Food and Agriculture Organization of the United Nations (FAO) Rome, Italy, 2017.

Pedgley, D. E. Testing Feasibility of Detecting Locust Breeding Sites by Satellite. Final Report to NASA on ERTS-1, Experiment, COPR, London. 1973.

Pekel, J. F.; Ceccato, P.; Vancutsem, C.; Cressman, K.; Vanbogaert, E.; Defourny, P. Development and Application of Multi-Temporal Colorimetric Transformation to Monitor Vegetation in the Desert Locust Habitat. *IEEE J. Sel. Top. Appl. Earth Obs. Remote Sens.* **2010,** *4* (2), 318–326.

Pener M. P.; Simpson S. J. Locust Phase Polyphenism: An Update. *Adv. Insect Physiol.* **2009,** *36*, 1–272.

Pener, M. P. Insect Growth Regulators. In *Encyclopedia of Environmental Management*, 4 Vol. Set; CRC Press, 2012; pp 1459–1470.

Peng, G.; Jin, K.; Liu, Y.; Xia, Y. Enhancing the Utilization of Host Trehalose by Fungal Trehalase Improves the Virulence of Fungal Insecticide. *Appl. Microbiol. Biotechnol.* **2015,** *99* (20), 8611–8618.

Peng, G.; Xia, Y. Expression of Scorpion Toxin Lqhit2 Increases the Virulence of *Metarhizium acridum* Towards *Locusta migratoria manilensis*. *J. Ind. Microbiol. Biotechnol.* **2014,** *41* (11), 1659–1666.

Peng, G.; Xia, Y. Integration of an Insecticidal Scorpion Toxin (Bjαit) Gene Into *Metarhizium acridum* Enhances Fungal Virulence Towards *Locusta migratoria manilensis*. *Pest Manag. Sci.* **2015,** *71* (1), 58–64.

Peng, G.; Xia, Y. The Mechanism of the Mycoinsecticide Diluent on the Efficacy of the Oil Formulation of Insecticidal Fungus. *BioControl* **2011,** *56* (6), 893–902.

Peng, W.; Ma, N. L.; Zhang, D.; Zhou, Q.; Yue, X.; Khoo, S. C.; Wang Y. A Review of Historical and Recent Locust Outbreaks: Links to Global Warming, Food Security and Mitigation Strategies. *Environ. Res*. **2020,** 110046.

Peshin, R. Integrated Pest Management: Pesticide Problems, vol. 3, 2014.

Phillips, D. R.; Loughton, B. G. The Effect of Inhibition of RNA and Protein Synthesis on the Development of Larvae of *Locusta migratoria*. *Insect Biochem*. **1979,** *9* (2), 241–245.

PikKheng, H.; Bong, C. F. J.; Jugah, K.; Rajan, A. Evaluation of *Metarhizium anisopliae* Var. *anisopliae* (Deuteromycotina: Hyphomycete) Isolates and Their Effects on Subterranean Termite *Coptotermes curvignathus* (Isoptera: Rhinotermitidae). *Am. J. Agric. Biol. Sci*. **2009,** *4* (4), 289–297.

Piou, C., Lebourgeois, V.; Benahi, A. S.; Bonnal, V.; el Hacen Jaavar, M.; Lecoq, M.; Vassal, J. M. Coupling Historical Prospection Data and a Remotely-Sensed Vegetation Index for the Preventative Control of Desert Locusts. *Basic Appl. Ecol*. **2013,** *14* (7), 593–604.

Prashar, P.; Shah, S. Impact Off Fertilizers and Pesticides on Soil Icroflora in Agriculture, Vol.12. 2013.

Pregitzer, P.; Jiang, X.; Grosse-Wilde, E.; Breer, H.; Krieger, J.; Fleischer, J. In Search for Pheromone Receptors: Certain Members of the Odorant Receptor Family in the Desert Locust *Schistocerca gregaria* (Orthoptera: Acrididae) Are Co-Expressed with SNMP1. *Int. J. Biol. Sci*. **2017,** *13* (7), 911–922.

Rajput, V. S.; Jhala, J.; Acharya, V. S. Biopesticides and Their Mode of Action Against Insect Pests: A Review. *IJCS.* **2020,** *8* (2), 2856–2862.

Ramesh, S. Biological and Environmental Hazards, Risks, and Disasters. Elsevier, Oxford, UK, 2015.

Rangel, D. E. N.; Braga, G. U. L.; Anderson, A. J.; Roberts, D. W. Influence of Growth Environment on Tolerance to UV-B Radiation, Germination Speed, and Morphology of *Metarhizium anisopliae* Var. *acridum* Conidia. *J. Invertebr. Pathol*. 2005, *90* (1), 55–58.

Razafindranaivo, V. Studies on the Nymphal Aggregation Pheromone of Malagasy Migratory Locust, *Locusta Migratoria Capito* (Saussure, 1884) and Its Effects on Adult Maturation, Ph.D. Thesis, Jomo Kenyatta University of Agriculture and Technology, Nairobi, Kenya, 2010.

Revankar, S. G.; Sutton, D. A.; Sanchez, S. E.; Rao, J.; Zervos, M.; Dashti, F.; Rinaldi, M. G. *Metarrhizium anisopliae* as a Cause of Sinusitis in Immunocompetent Hosts. *J. Clin. Microbiol*. **1999,** *37* (1), 195–198.

Ripley, S. H.; Ewer, D. W. Peripheral Inhibition in the Locust. *Nature* **1951,** *167* (4261), 1066–1067.

Rosenberg, L. J. Information Systems for Locust Forecasting. In *Research Priorities for Migrant Pests in Southern Africa*; Cheke, R. A., Rosenberg, L. J., Kieser, M. E., Eds.; Natural Resources Institute: London, Great Britain, **2000;** pp 165–172.

Santos, D., Broeck, J. V.; Wynant, N. Systemic RNA Interference in Locusts: Reverse Genetics and Possibilities for Locust Pest Control. *Curr. Opin. Insect Sci*. 2014, *6*, 9–14.

Scherer, R. Rakotonandrasana MA. Barrier Treatment with a Benzoyl Urea Insect Growth Regulator Against *Locusta migratoria capito* (Sauss) Hopper Bands in Madagascar. *Int. J. Pest Manage*. **1993,** *39* (4), 411–417.

Schiel, L.; Wind, C.; Braun, P. G.; Koethe, M. Legal Framework for the Marketing of Food Insects in the European Union. Ernahrungs Um-schau. **2020,** *67* (4), 76–85.

Seyoum, E.; Negash, M. Studies on the Field Performance of *Metarhizium Anisoliae* Var. *Acridum* (Green Muscle) Against Mixed Populations of Grasshopper in Ethiopia. *SINET Ethiop. J. Sci.* **2007,** *30* (1), 55–64.

Sh, H. K.; Ghoneim, K. S.; El-Hela, A. A.; Amer, S. M.; Mohammad, A. A. Disturbed Survival, Growth and Development of the Desert Locust *Schistocerca Gregaria* by Different Extracts of *Azadirachta indica* (Meliaceae) and *Nigella sativa* (Ranunculaceae). *Egypt. Acad. J. Biol. Sci. A, Entomol.* **2013,** *6* (2), 1–21.

Shah, F. A.; Ansari, M. A.; Watkins, J.; Phelps, Z.; Cross, J.; Butt, T. M. Influence of Commercial Fungicides on the Germination, Growth and Virulence of Four Species of Entomopathogenic Fungi. *Biocontrol Sci. Tech.* **2009,** *19* (7), 743–753.

Sharmila, D. Locust Swarms in East Africa Could be "A Catastrophe". *Lancet* 2020, *395* (10224), 547.

Shi, R. X.; Liu, C.; Li, D. M.; Xie, B. Y. Application of MODIS_NDVI to Monitoring Locust Pest in Baiyang Shallow Lake. *J. Nat. Disasters*. **2003,** *12* (3), 155–160.

Shi, W.; Guo, Y.; Xu, C.; Tan, S.; Miao, J.; Feng, Y.; Zhao, H.; Leger, R. J. S.; Fang, W. Unveiling the Mechanism by Which Microsporidian Parasites Prevent Locust Swarm Behavior. *Proc. Natl. Acad. Sci. USA*. **2014,** *111* (4), 1343–1348.

Showail, M. J.; Kus, J. V.; Tsui, G. K.; Chew, H. F. Fungal Keratitis Caused by *Metarhizium anisopliae* Complex. *Med. Mycol. Case Rep.* **2017,** *17*, 28–30.

Siegel, J. P. Testing the Pathogenicity and Infectivity of Entomopathogens to Mammals. In *Manual of Techniques in Insect Pathology*; Lacey, L., Ed.; Academic Press: San Diego, CA, **1997;** pp 325–336.

Simpson, S. J.; Sword, G. A. Locusts. *Curr. Biol.* 2008, *18* (9), R364–366.

Singh, R. P., Pant, N. C. Lycorine-A Resistance Factor in the Plants of Subfamily Amaryllidoideae (Amaryllidaceae) Against Desert Locust, *Schistocerca gregaria* F. *Experientia* **1980,** *36* (5), 552–553.

Singh, S.; Pandher, S.; Sharma, R. K.; Kumar, R. Insect Growth Regulators: Practical Use, Limitations and Future. *J. Eco-friendly Agric.* **2013,** *8* (1), 1–4.

Sivanpillai, R.; Latchininsky, A. V. Mapping Locust Habitats in Amudarya River Delta, Uzbekistan with Multi-Temporal MODIS Imagery. *Environ. Manage.* **2007,** *39* (6), 876–886.

Sivanpillai, R.; Latchininsky, A. V.; Driese, K. L.; Kambulin, V. E. Mapping Locust Habitats in River Ili Delta, Kazakhstan, Using Landsat Imagery. *Agric. Ecosyst. Environ.* **2006,** *117* (2–3), 128–134.

Sivanpillai, R.; Latchininsky, A. V.; Peveling, R.; Pankov, V. I. Utility of the IRS-Awifs Data to Map the Potential Italian Locust (*Calliptamus Italicus*) Habitats in Northeast Kazakhstan. *Proceed. 2009 ASPRS Annu. Conf.*, ASPRS: Baltimore, MD. 2009.

Smith, R. F.; van den Bosch, R. Integrated Control. In *Pest Control: Biological, Physical, and Selected Chemical Methods*; Kilgore, W. W., Doutt, R. L. Eds.; Academic Press: New York **1967;** pp 295–340.

Song, H. Density-Dependent Phase Polyphenism in Nonmodel Locusts: A Minireview. *Psyche* **2011,** *16* (2011).

Song, L.; Gao, M.; Dai, S.; Wu, Y.; Yi, D.; Li, R. Specific Activity of a *Bacillus Thuringiensis* Strain Against *Locusta migratoria manilensis*. *J. Invertebr. Pathol.* **2008,** *98* (2), 169–175.

Song. Q.; Zhang, Y.; Yan, L.; Wang, J.; Lu, C.; Zhang, Q.; Zhao, M. Risk Assessment of the Endocrine-Disrupting Effects of Nine Chiral Pesticides. **2017,** *J. Hazard Mater. 338,* 57–65.

Stahr, C.; Seidelmann, K. Individual Pheromone Signature in Males: Prerequisite for Pheromone-Mediated Mate Assessment in the Central American Locust, *Schistocerca piceifrons*. *J. Chem. Ecol.* **2016,** *42* (12), 1304–1313.

Starnes, R. L.; Liu, C. L.; Marroue, P. G. History, Use and Future of Microbial Insecticides. *Am. Entomol.* **1993,** *39* (2), 83–91.

Steedman, A. *Locust Handbook,* 2nd ed.; Overseas Development Natural Resource Institute: London, 1988.

Story. P. G.; Walker, P. W.; McRae, H.; Hamilton, J. G. A Case Study of the Australian Plague Locust Commission and Environmental Due Diligence: Why Mere Legislative Compliance Is No Longer Sufficient for Environmentally Responsible Locust Control in Australia. Integr. Environ. Assess. Manag. an Int. J. **2005,** *1* (3), 245–251.

Tan, S. Q.; Yin, Y.; Cao, K. L.; Zhao, X. X.; Wang, X. Y.; Zhang, Y. X.; Shi, W. P. Effects of a Combined Infection with *Paranosema locustae* and *Beauveria bassiana* on *Locusta migratoria* and Its Gut Microflora. *Insect Sci*. Epub Ahead of Print 13 March 2020.

Tanaka, S.; Kotaki, T.; Nishide, Y.; Ben-Hamouda, A.; Abdellaoui, K.; Ebbe, M. A.; Ely, S. O. Effects of Water Extracts of Frass from Three Locust Species and Various Plants on Oviposition and Embryonic Development in the Desert Locust, *Schistocerca gregaria*. *J. Orthoptera Res*. **2019,** *28*, 195.

Tanani, M. Dual Effect of the Antibiotic Cycloheximide on Desert Locust *Schistocerca gregaria* Forskal (Orthoptera: Acrididae). *J. Entomol. Zool. Stud.* 2018, *6* (5), 538–548.

Tao, K.; Long, Z.; Liu, K.; Tao, Y.; Liu, S. Purification and Properties of a Novel Insecticidal Protein from the Locust Pathogen *Serratia marcescens* HR-3. *Curr. Microbial*. **2006,** *52* (1), 45–49.

Tao, K.; Yu, X.; Liu, Y.; Shi, G.; Liu, S.; Hou, T. Cloning, Expression, and Purification of Insecticidal Protein Pr596 from Locust Pathogen *Serratia marcescens* HR-3. *Curr. Microbiol.* **2007,** *55* (3), 228–233.

Tian, H. D.; Ji, R.; Xie, B. Y.; Li, X. H.; Li, D. M. Using Multi–Temporal Landsat ETM+ Data to Monitor the Plague of Oriental Migratory Locust. *Int. J. Remote Sens*. **2008,** *29* (6), 1685–1692.

Tkaczuk, C.; Majchrowska-Safaryan, A.; Miętkiewski, R. The Influence of Selected Fungicides and Soil Extracts on the Growth of Entomopathogenic Fungus *Metarhizium anisopliae*. *Prog. Plant Protec*. **2013,** *53* (4).

Toriello, C.; Navarro-Barranco, H.; Martínez, J. A.; Mier, T. Seguridad En Ratones De *Metarhizium anisopliae* (Metsch.) Sorokin Aislado De *Aeneolamia* Sp. (Homoptera: Cercopidae). *Rev. Mex. Micol*. **1999,** *3* (15), 123–125.

Toriello, C.; Pérez-Torres, A.; Burciaga-Díaz, A.; Navarro-Barranco, H.; Pérez-Mejía, A.; Lorenzana-Jiménez, M.; Mier, T. Lack of Acute Pathogenicity and Toxicity in Mice of an Isolated of *Metarhizium anisopliae* Var. *anisopliae* from Spittlebugs. *Ecotoxicol. Environ. Saf.* **2006,** *65* (2), 278–287.

Toriello, C.; Pérez-Torres, A.; Vega-García, F.; Navarro-Barranco, H.; Pérez-Mejía, A.; Lorenzana-Jiménez, M.; Hernández-Velázquez, V; Mier, T. Lack of Pathogenicity and Toxicity of the Mycoinsecticide *Metarhizium anisopliae* var. *acridum* Following Acute Gastric Exposure in Mice. *Ecotoxicol. Environ. Saf.* **2009,** *72* (8), 2153–2157.

Tounou, A. K.; Kooyman, C.; Douro-Kpindou, O. K.; Poehling, H. M. Interaction Between *Paranosema Locustae* and *Metarhizium Anisopliae* Var. *Acridum*, Two Pathogens of the Desert Locust, *Schistocerca gregaria* Under Laboratory Conditions. *J. Invertebr. Pathol.* **2008,** *97* (3), 203–210.

Tucker, C. J.; Hielkema, J. U.; Roffey, J. The Potential of Satellite Remote Sensing of Ecological Conditions for Survey and Forecasting Desert Locust Activity. *Int. J. Remote Sens*. 1985, *6* (1), 127–138.

US National Academy of Sciences, 1969. Insect Pest Management and Control. Washington, D.C: National Academy of Sciences Publication 1695.

Usmanov, S.; Gapparov, F. Biological Efficiency of New Insecticides Against Harmful Locusts in Uzbekistan. *Am. J. Appl. Sci*. **2020,** *2* (9), 299–303.

Uvarov, B. P. The Locust Plague. *J. Econ. Entomol*. **1944,** *37* (1), 93–99.

Valizadeh, H.; Abbasipour, H.; Mahmoudvand, M.; Askary, H.; Moniri, V. R. Laboratory Trials of *Metarhizium Anisopliae* Var. *Acridum* (Green Muscle®) Against the Saxaul Locust, *Dericorys albidula serville* (Orthoptera: Dericorythidae). *Chil. J. Agric. Res.* **2011,** *71* (4), 549–553.

Valverde-Garcia, P.; Santiago-Álvarez, C.; Thomas, M. B.; Maranhao, E. A.; Garrido-Jurado, I.; Quesada-Moraga, E. Sublethal Effects of Mixed Fungal Infections on the Moroccan Locust, *Dociostaurus maroccanus*. *J. Invertebr. Pathol*. **2019,** *161*, 61–69.

Van Huis A. Strategies to Control the Desert Locust *Schistocerca gregaria* in Area-Wide Control of Insect Pests Springer, Dordrecht. **2007,** 285–296.

Vega, F. E.; Meyling, N. V.; Luangsa-Ard, J. J. Fungal Entomopathogens. In *Insect Pathology*; Vega, F. E., Kaya H. K., Eds., 2nd ed.; Academic Press: New York, **2012;** pp 171–122.

Verlinden, H.; Sterck, L.; Li, J.; Li, Z.; Yssel, A.; Gansemans, Y.; Verdonck, R.; Holtof, M.; Song, H.; Behmer. S. T.; Sword, G. A. First Draft Genome Assembly of the Desert Locust, *Schistocerca gregaria. F1000Res*. **2020,** *9* (775), 775.

Vijayalakshmia, A.; Meenaa, M. Crop Destructor-Locust-A Critical Review. *J. Xidian Univ*. **2020,** *14* (7), 1798–1804.

Vincent, C.; Goettel, M. S.; Lazarovitis, G. Biological Control: A Global Perspective: Case Studies from Around the World. (Cabi, Oxfordshire, 2007).

Voss, F.; Dreiser, U. Mapping of Desert Locust Habitats Using Remote Sensing Techniques. In *New Strategies in Locust Control*; Krall, S.; Peveling, R.; Ba Diallo, D., Eds.; Birkhäuser: Basel, Switzerland, **1997;** pp 37–45.

Vreysen, M. J.; Robinson, A. S.; Hendrichs, J.; Kenmore, P. Area-Wide Integrated Pest Management (AW-IPM): Principles, Practice and Prospects. In Area-*Wide Control of Insect Pests*; Springer: Dordrecht, **2007;** pp 3–33.

Wahed, A. E.; Samira, M. N.; Elhadidy, N. M. Immunity Changes in *Locusta Migratoria* Linnaeus (Orthoptera: Acrididae) Infected by Entomopathogenic Nematode *Steinernema carpocapsae* (Rhabditida: Steinernematidae). *J. Plant Protec. Pathol*. **2018,** *9* (12), 877–881.

Wakgari, W. A. Comparison of Efficacies of Bran-Based Baits Containing Diflubenzuron, Teflubenzuron, and Fenitrothion Against Desert Locust, *Schistocerca gregaria* (Forsk.) (Orthoptera: Acrididae). *Int. J. Pest Manage*. **1997,** *43* (2), 163–167.

Walsh, J. Return of the Locust: A Cloud Over Africa: Last Year's Rains Brought Better Harvests–And a Bigger Crop of Locusts and, Grasshoppers, Threatening a Revival of the Plagues of the Past. *Science* **1986,** *234* (4772), 17–19.

Wang, B.; Kang, Q.; Lu, Y.; Bai, L.; Wang, C. Unveiling the Biosynthetic Puzzle of Destruxins in *Metarhizium* Species. *Proceed. Natl. Acad. Sci.* 2012, *109* (4), 1287–1292.

Wang, F.; F. Schnal. Effects of the Imidazole Derivative KK-42 on the Female and Embryos of *Schistocerca gregaria*. *Entomol. Sci.* **2001,** *4* (4), 387–392.

Wang, H. S.; Wang, X. H.; Zhou, C. S.; Huang, L. H.; Zhang, S. F.; Guo, W.; Kang, L. Cdna Cloning of Heat Shock Proteins and Their Expression in the Two Phases of the Migratory Locust. *Insect Mol. Biol.* **2007,** *16* (2), 207–219.

Wang, L. X.; Zhou, Q. Nepenthes Pitchers: Surface Structure, Physical Property, Anti-Attachment Function and Potential Application in Mechanical Controlling Plague Locust. *Chin. Sci. Bull.* **2014,** *59* (21), 2513–2523.

Wang, S.; Fang, W.; Wang, C.; St. Leger, R. J. Insertion of an Esterase Gene Into a Specific Locust Pathogen (*Metarhizium Acridum*) Enables It to Infect Caterpillars. *PLoS Pathog.* **2011,** *7* (6), e1002097.

Whitten, M. J.; Oakeshott, J. G. Opportunities for Modern Biotechnology in Control of Insect Pests and Weeds, with Special Reference to Developing Countries. *FAO Plant Protec. Bull.* (Multilingual ed.). **1991,** *39* (4), 155–181.

Wu, Y.; Lei, C. F.; Yi, D.; Liu, P. M.; Gao, M. Y. Novel *Bacillus Thuringiensis* Δ-Endotoxin Active Against *Locusta migratoria manilensis*. *Appl. Environ. Microbiol.* **2011,** *77* (10), 3227–3233.

Xu, R. Q.; Zhou, Q.; Wang, S. M. Development of Mechanized Capture and Control Technology of Locust Disaster. *J. Agric. Mac.* **2005,** *36* (11), 171–173.

Yu, G.; Shen, H.; Liu, J. Impacts of Climate Change on Historical Locust Outbreaks in China. *J. Geophys. Res.* **2009,** *114* (D18), D18104.

Zaim, A.; El-Ghadraoui, L.; Farah, A. Effects of *Artemisia herba-alba* Essential Oils on the Survival of Adult Locusts of *Euchorthippus albolineatus* (Lucas, 1849). Bulletin De l'Institut Scientifique: Section Sciences De La Vie. **2012,** *34* (2), 127–133.

Zhang, L.; Lecoq, M.; Latchininsky, A.; Hunter, D. Locust and Grasshopper Management. *Annu. Rev. Entomol.* **2019,** *64*, 15–34.

Zhang, P. F.; You, Y. W.; Song, Y.; Wang, Y. Z.; Zhang, L. First Record of *Aspergillus Oryzae* (Eurotiales: Trichocomaceae) as an Entomopathogenic Fungus of the Locust, *Locusta migratoria* (Orthoptera: Acrididae). *Biocontrol Sci. Technol.* **2015,** *25* (11), 1285–1298.

Zhiwei, Z.; Xiaohui, Z.; Yusheng, L. Present Situation of Sucking Pneumatic Techniques for Capturing Locusts and Utilization of Artificial Raising Locusts. *J. Agric. Eng.* **2007,** *23*, 126–131.

Zimmermann, G. Review on Safety of the Entomopathogenic Fungus *Metarhizium anisopliae*. *Biocontrol Sci. Technol.* **2007,** *17* (9), 879–920.

Zimmermann, G. The Entomopathogenic Fungus *Metarhizium anisopliae* and Its Potential as a Biocontrol Agent. *J. Pest. Sci.* **1993,** *37* (4), 375–379.

CHAPTER 5

Advanced Technologies for Monitoring and Management of Locusts

MIRZA ABDUL QAYYUM[1*], MUHAMMAD YASIN[2], WAQAS WAKIL[3,4], DAVID HUNTER[5], M. USMAN GHAZANFAR[6], MUHAMMAD WAJID[1], SHAFQAT SAEED[1], and MUHAMMAD ASHFAQ[1]

[1]*Institute of Plant Protection, MNS University of Agriculture, Multan*

[2]*Department of Entomology, The Islamia University, Bahawalpur*

[3]*Department of Entomology, University of Agriculture, Faisalabad, Pakistan*

[4]*Department of Continuing Education, University of Agriculture, Faisalabad, Pakistan*

[5]*Orthopterists' Society, McKellar, Australia*

[6]*College of Agriculture, University of Sargodha, Pakistan*

ABSTRACT

A dozen locust species are threatening food security worldwide. Despite the enormous research in exploring locust ecological behavior, exact prediction, and management, locust outbreaks are still found insufficient, as outbreaks of locust are still occurring during the 21st century. Locusts typically inhabit and reside at remote and scarcely dense vegetation areas, and are ultimately distributed to ranges across different continents.

Locust Outbreaks: Management and the World Economy. Umair Riaz, Khalid Rehman Hakeem, (Eds.)

This inappropriate knowledge has a tremendous impact on insufficient monitoring and management models to avoid locust outbreaks. In the recent past, traditional surveys on locusts found it unsatisfactory to cover the wider area as required. The use of remote sensing in monitoring locust outbreaks proved a promising tool. Using satellite-based data for monitoring and forecasting locust species is in the right business. Unmanned aerial vehicles or simply drones are an innovative technology for monitoring locusts. Forecasting is another technique to examine the invasive nature of the desert locust under changing field conditions. Assessing the environmental condition is a more important tool as it indicates a conducive environment for breeding. In recent years, FAO DLIS reported and adopted drone technology for early detection and prevention by finding the accurate green patch that is a relatively more suited place for the locust infestation. Drones are also used to detect post-disaster mapping to assess the damage caused by desert locusts. The government and the competent authorities need data about the extent of damage afterward. Drones are replacing the more expensive light aircraft equipped with real-time thermal and image processing sensors to support satellite imagery systems. eLocust is another advanced tool for locust monitoring that is used for recording and transmission of locust data using electronic devices with custom software. The prime objective is to view and navigate the swarms along with the location. The data collected from this include locust stages, habitat type, and vegetation species, the treatment used, and safety precautions that can be integrated with data for rainfall and green vegetations. Reconnaissance and Monitoring System of the Environment of Schistocerca (RAMSE Sv4 (2015)) is one more revolution by the FAO used in countries where locust outbreaks are known to occur to explore the biology and ecology for better management of locust populations. It also works on the data generated by the eLocust3, which requires open-source Global Information Systems Software, that is, OpenJump and Post GIS PostGresSQL. Schistocerca Warning and Management System (SWARMS) is a technological tool widely used for better integration, display, comparison, and real-time maps at all appropriate data units to forecast and predict locust populations across international levels. As far as the intensity of the locust plague is concerned, the only way the world can come up is by equipping itself with the latest technologies and strategies.

5.1 BACKGROUND

Locust and grasshoppers are the most devastating threats to agro-ecosystems worldwide. Locust alone comprises hundreds of species that affect farmers' livelihood from one in every tenth farmer worldwide (Latchininsky et al., 2011). The recorded species of Acrididae alone consists of more than 500 species worldwide, which are damaging the green pastures and many crops (Zhang et al., 2019). By using better management and early detection protocols, locust outbreaks were not only controlled but also the invasion potential will be reduced to some extent. But large-scale invasions still continue to occur in many countries worldwide. Recently, the outbreak of desert locust covering the larger part of the African peninsula and southwest Asia from 1986 to 2016 covered about 2.3 million hectares of agricultural land and 37 million US dollar were being used to manage this pest (FAO, 2018). Approximately, more than 1.5 million hectares of land are invaded and ultimately infested annually reducing crop production as well as economic stability (Yang and Ren, 2018). While studying the locust swarming behavior and its physiology, the population is increasing at an exponential rate that may lead to catastrophic ecological impact devastating food security and socioeconomic aspects (Crook et al., 2020). This has led to increasing the challenges to prevent its invasion and control tactics for better management (Yao et al., 2017).

TABLE 5.1 Desert Outbreaks Episodes and Control.

Locust episodes	Countries	Amount spent	Duration	Cost of donors	Intervention
1986–1989	23	25 million	4 years	$310 million	Late
1992–1994	18	4 million	2 years	$18.8 million	moderate
1997–1998	7	4.3 million	7 months	$30,000	Early
2003–2005	20	13 million	2.5 years	$35 million	Late
2007–2016	22	1.8million	10 years	<$1 million	Early

Source: Reprinted from Showler et al. (2021).

Traditional approaches in locust control were based only on artificial investigation for monitoring purposes; that was inefficient, less accurate, and not useful to prevent its invasion. As locust outbreaks were distributed to extensive areas, modern technologies based on satellites and remote sensing proved very useful for monitoring and conceiving the idea about locust habitat worldwide (Mohammed et al., 2015). Chemical use was the only solution for the management of locust outbreaks with serious environmental hazards. These include water pollution, biodiversity reduction, bioaccumulation, and ecological damage (Singh et al., 2018). For sustainable agriculture and pest management, minimum use of insecticides is recommended which is not possible in locust control cases (Bais-Moleman et al., 2019). So, for better ecological control and effective management techniques, one should rely on modern strategies based on the latest emerging techniques for locust control programs. Some of them that are very useful are given as under.

5.2 LAND USE AND COVER CHANGE

Land Use and Cover Change is an important and updated cartography; it is one of the most evident and fast-changing processes for both land surface and also natural ecosystem services (Herold et al., 2009). LUCC is an important factor in driving and assessing changes in the environment worldwide (Lawrence and Chase, 2010). It is also the indicator to understand the comparative interaction of human activities with its environment (Dewan et al., 2012). The basic objective of LUCC is to enable the scientists to pick the most important systematic transitions in grasslands or agro-ecological zones for better assessments of these proximate reasons of change.

In the desert locust management program, the formation, and dynamics of changes in locust plaque areas are directly linked with LUCC. Clearing the LUCC is very vigilant in implementing the rational as well as ecological control programs in policymaking and timely early detection of potential locust areas. LUCC provides us with the essential information about the outbreak area and its further extraction. The dynamic changes in the locust area will change as its habitat changes which are assessed by LUCC. It helps in quantitative analysis and the evolutionary changes in locust areas to evaluate the benchmark of ecological control measures. It may give insights into the timely detection of locusts over large and far inaccessible areas of potential plaque. The remote sensing technology along with LUCC helps to capture data along with monitoring and qualitative analysis. There are many inconveniences while collecting data and analysis of land cover data because of no uniform standards in land use classification.

To find out the relationship between LUCC and areas of plaque, the land Use Transfer Matrix (LUTM) is used which indicates the area of LUC type in a period. Then there are two indicators, the Transition Probability Index (TPI) and the Proportional Differential Index (PDI).

5.3 NORMALIZED DIFFERENCE VEGETATION INDEX (NDVI)

The vegetation index is leading to managing desert locust as it may solve the spatiotemporal hurdles in decision-making whether to survey or not. NDVI or vegetation greenness maps using remote sensing helped the scientist to indicate areas where locust populations may develop soon. To decide to survey, it is very necessary to have information about moisture contents in soil which ultimately grow vegetation. Apart from that, soil moisture is also helpful in indicating the potential areas of locust where it lays eggs, preferably in moist soils (Penner and Simpson, 2009). Lazar et al (2016) revealed that areas where heavy rainfall occurred, or mountainous areas favor the development of this vegetation and suitable indicators of locust reproduction over large areas.

Vegetation cover supports the locust population because of feeding and shelter source. So, it is also helpful in monitoring the locust population that is distributed to many vegetation areas. Areas of low vegetation cover support oviposition and high cover deter (Han, 2003). In the areas of about 90% vegetation, adult and nymph populations are unable to migrate (Ma et al., 2007). Therefore, these are the main parameters to formulate monitoring models to derive infestation rates from areas of changing vegetation by NDVI (Shen and Shen, 2005). So, effective use of satellite and remote sensing is used to detect the vegetation cover using the NDVI technique, distinguishing the areas in bare soils from green soils (Cressman, 2013). FAO DLIS (Desert Locust Information System) is utilizing the MODIS (Moderate Resolution Imaging Spectroradiometer) for a spatial resolution to generate NDVI values (Tratalos and Cheke, 2003). Thus, MODIS data provides Greenness maps from the NDVI index (Renier et al., 2015).

5.4 THE MODERATE RESOLUTION IMAGE SPECTRORADIOMETER

Nowadays, locust surveys include practices that work more accurately and sophisticatedly spatiotemporal information about habitat distributions. So, satellite-based observations using the Moderate Resolution

Image Spectroradiometer satellites are used for both early detection and monitoring of locust populations that could infest potential areas during a particular growing season (Low et al., 2016). Most spectral data based on different satellites gave data that is useful in seasonal as well-off seasonal monitoring of locust habitat (Sivanpillai and Latchininsky, 2007). MODIS is used, for instance, to get spectral information based on thermal and infrared radiations-making images of potential swarms at study locations. The first step in MODIS includes the segmentation of objects oriented to different lands to get homogeneous objects (Hüttich et al., 2009). There are many satellites and sensors based on MODIS that are used to monitor desert locust populations including Aqua MODIS, Terra MODIS, Meteosat-8, and Landsat series (Gómez et al., 2019).

Zhang et al (2019) studied and assessed locust habitat using MODIS on different growth stages of locust. The primary data generated from his study was oviposition, incubation, and development of locust using Oviposition, incubation, and development suitability indicators. The complete coverage using MODIS was well studied by Shen and Shen (2005) to evaluate the relationship between vegetation and vegetations consumed by locust populations. The results helped them to conclude that NDVI values show vegetation growth suppression when locusts are at the nymph stage.

Temperature vegetation dryness index data using MODIS was studied by Crook and Robinson (2019) to evaluate the oviposition period and incubation period of locust swarms at three different regions. They have extracted NDVI data from MODIS, got data about greenness maps during plague year, and the extent of locust hatching in potential areas. The Desert Locust Information Service (DLIS) working under Food and Agriculture Organization (FAO) uses MODIS to generate images at different resolutions of around 250 meters and got NDVI values of more than 16-day composite images (Tratalos and Cheke, 2009). The data from MODIS will be exploited with dynamic greenness maps using the NDVI index (Renier et al., 2015). Deveson (2013) in Australia used MODIS-derived NDVI to forecast locust populations.

5.5 RNAi

From the last few decades, enormous progress has been made on the potential use of RNA interface (commonly written as RNAi) to control pest populations in agronomic as well as horticultural crops.

The efficient use in insect pest management using RNAi developed certain novel approaches. The RNAi technique is widely understood as a genetic tool to knock down the function of genes. The RNAi contributed to newer strategies for controlling insect pest species crops (Whyard et al., 2009). dsRNA-mediated gene silencing has many essential implications and possibilities using silencing techniques in the locust control program (Joga et al., 2016). Systemic use of RNAi uses dsRNA uptake method using injecting, adsorbing, or feeding by the insect body and ultimately persistent suppression from the target gene of messenger RNA. The dsRNA utilized from the external to the internal application using RNAi among cells and tissues in the locust body was studied. It is a strategy regarding the molecular mechanism between the uptake and transformation of silencing genes (Zotti and Smagghe, 2015).

Gijbels et al (2020) reported the major effects using RNAi-mediated SgMet knockdown in female desert locust *S. gregaria.* Juvenile hormone contributes to the development of managerial and target-based insecticides to kill voracious pests like the locust. By the use of the RNAi technique, they utilized the JH receptors that are essential in ovarian development in the female desert locust body for adultery. Knockdown of *SgMet* results in reduced production of insulin-related peptides (SgIRP) and enhanced production of Neuroparsin (SgNP) in the insect body. This technique regulates the hormonal imbalance of the insect body that ultimately results in ineffective development phases.

The molecular and developmental effects due to RNAi knockdown of SgE93 hormone in desert locust were studied by Gijbels et al (2020). The effects shown after the injection of dsRNA based on SgE93 inhibit adult metamorphosis instead of remaining stuck in the nymph stage. After comparing to control the locust population, a slightly shorter final nymph stage was recorded. RNases presence in any insect body usually in the midgut made it unmanageable using RNAi. The desert locust expressed dsRNases in the midgut (Wynant et al., 2014). The variation among different geographical populations of desert locusts using the RNAi technique was also studied. But it resulted in no clear evidence about sensitivity using RNAi. The test insects were two strains from different geographical areas and were found sensitive using dsRNA. Corazonin (CRZ) and Ecdysone receptor hormones were found resistant (Sugahara et al., 2017).

5.6 GIS AND GPS

It was the era of the early 1990s when the Geographical Information system was introduced that helped the scientific community in geographical mappings. GIS consists of two components including a database and application software for mapping. Latitude and longitude based must be referenced to assure accuracy on maps. In 1994, the dreams came true when the Desert Locust Information System (DLIS) along with the Natural Resources Institute and University of Edinburgh (Geography department) developed GIS to monitor and early detect the desert locust (Klein et al., 2021). The GIS system allows scientists to access large data based on weather data and locust populations that can predict future swarms of the desert locust. Before the invention of GPS and GIS technology, researchers found it difficult to locate the position of using paper maps and to plot infestations; one should use overlays using a lead pencil.

GIS and GPS are generally an information support system that contains data of locust populations and habitat suitability to forecast future swarms (Sivanpillai et al., 2006). With the advent of GIS and GPS, the analysis of composite spatiotemporal patterns made it easy for all scientists including entomologists and ecologists. It relates locust outbreaks with habitat dynamics affected by locusts. Both patterns are mapped and integrated into the GIS system (Cigliano and Torrusio, 2017). Apart from the above-mentioned discussion, spatial analysis based on locust distributions can be located (Yao et al., 2014). Scientists can locate probable breeding sites (Mohammed et al., 2015) and also analyze the spatiotemporal dynamics (Veran et al., 2015).

Currently, many GIS applications are made for locust control programs based on web-based servers restricted to limited users, large data sources, and sharing systems. WebGIS is an example of a GIS system that is a potential application tool with almost minimum resources that overcomes the extra burden to collect data, save time, and is easily communicable to other scientific communities (Silva et al., 2014).

5.7 DRONES

Drones being an innovative technology are widely used in agriculture to optimize agricultural operations, enhancing crop productivity, and monitoring plant growth. Monitoring includes the use of sensors and imaging of fields which can assist in different ways including crop growth and productivity. It is used for both surveying and management of desert locusts. In this regard space technology is used for the sake of monitoring the breeding and swarming. Forecasting is another technique to monitor the invasion biology of the desert locust. Assessing the environmental condition is a more important tool as it indicates a conducive environment for breeding. Spatiotemporal data may help in the distribution biology of the desert locust.

In recent past years, FAO DLIS reported and adopted drone technology for early detection and prevention to address the current challenges. From this, they succeeded in finding the accurate green patch that is a relatively more suited place for the locust infestation. Two types of drones are used: the first one is fixed-wing for the surveillance of infested areas and the second one is a rotary-wing drone that takes real-time pictures of the vegetation to assess the exact concentration of the locust group. A special operation like spot spraying is done to avoid pesticide drift and operational contaminations.

- Limited carrying capacity is a serious issue as it is experienced by the different research for optimized spray. Using drones carrying insecticides covers inadequate distances due to the size of the drone.
- Many researchers are doing work to make such prototypes that can effectively detect locust swarms using sensors, adapted to carry much volume to spray and cover large distances. Legislation is another challenging thing in many countries to operate drones because many countries are deploying laws and prohibiting flights under exceptional circumstances. So, larger drones with carrying capacities of 1500 liters and long traveling distances require special permissions.
- Drones are also used to detect post-disaster mapping to assess the damage caused by desert locusts. The government and the competent authorities need data about the extent of damage afterward. These high-resolution images give more accurate assessments of crop loss and health.
- In Australia, drones are replacing the more expensive light aircraft that are used for remote sensing and carrying data about surveillance. These drones are equipped with real-time thermal and image processing sensors to support the satellite imagery systems.

5.8 RECONNAISSANCE AND MONITORING SYSTEM OF THE ENVIRONMENT OF SCHISTROCERA (RAMSES)

A system devised by the FAO is widely used in areas with high locust populations to investigate and ultimately understand the locust ecology for better management. It also works on the data generated by the eLocust3. It also uses open-source Global Information Systems Software like Open-Jump and PostGISPostGresSQL. The data is also transmitted to FAO headquarters for analysis and storage. RAMSES allows NLI services to generate, manage, and analyze recorded data to map the entire field-based data (Cressman, 2013). All the field data from field surveys, reports, and management tactics along with geo-referenced GPS are downloaded on personal computers that will be transmitted by the satellites via the Internet to FAO headquarter in real time. The stored data is then analyzed on different national and international organizations for the analysis of desert locust distributions in diverse geographical regions around the globe (Cressman, 2016). Firstly, the RAMSES was introduced in 2000, which was operated using personal computers via Microsoft Access database and ArcView software. In 2014, some open source-based platform versions were made using independent software like OpenJump GIS and PostGress spatial databases.

5.9 SCHISTOCERA WARNING AND MANAGEMENT SYSTEM (SWARMS)

DLIS used SWARMS in 1996 for the management and analysis of locust and environmental data. It is basically the first GIS system dully used for operational monitoring of locust swarms rather than for mapping. It is the only GIS system using an Oracle database that collects information about surveys and data from various geographical areas where desert locusts attack. It is also used for remote sensing to save images along with ESRI's ArcGIS software for graphical display and spatial analysis. It is also a server-based system supported by a PC that allows scientists to access all the data collected. It is a program widely used for integration, comparison, and showcases all about locust to forecast at any level. It is used for the combined analysis of the spatial relationship between infestations.

- We can conclude that drone technology has proved an innovation being tried in the world in crop protection, to assess crop production and postdisaster aftermaths. Also, the software given above is helpful for early detection and management techniques. It will also give encouraging results if applied in Pakistan to combat locusts.

5.10 ELOCUST (FROM ELOCUST2 TO ELOCUST3)

A system is used for recording and transmission of locust data using electronic devices with custom software. The prime objective is to view and navigate the swarms along with the location. The data is transmitted via satellite for early warning systems and preventive efforts by the respective authorities. The data collected from this include locust stages, habitat type, and vegetation species, treatment used, and safety precautions. The data is combined to assess weather patterns including rainfall, temperature, relative humidity, and dense green vegetation based on satellite images, chronological records, and other developments and models to analyze the spread of the desert. Data forms and data collected from RAMSES will be transmitted to DLIS via email. Compared to older techniques, a strong and more reliable information system is developed that keeps operating between locust-infected countries of the world and DLIS.

In 2000, a prototype was developed to demonstrate the data based on surveys that were digitally entered into the database in the field by locust control officers. This eLocust technology consists of a handheld computer (Psion 5mx palmtop) and a database linked with mapping application software. A GPS device is there which is connected with portable alkaline batteries for power. After some years, with the help of Novacom Services elocust2 was developed that is equipped with a data logger having the touch screen, custom software in both English and French language, and an antenna for positioning of field operations. Real-time data transmission is done using the Inmarsat satellite.

In 2014, eLocust2 was replaced with the development of eLocust3 having the advantage of the latest technologies. Relatively smaller in size, it consists of TouchPad having applications to enter data, to view Landsat images, weather data, internet connectivity, wireless facility for the transmission of data, and a camera.

5.11 AMPLIFIED LOCUST SEARCH

ALS is an optimization algorithm that is used to explore locusts search with altered positions of locusts using algorithms in its evolution. There are two behavioral phases in desert locusts; one is the solitary phase and the second is a social or gregarious phase. This algorithm is an alternative to apply both the solitary as well as gregarious phase (Kanagasabai, 2020).

5.12 CONCLUSIONS

From the early 1990s to now 2020, seven large locust outbreaks have hit the world that affects the agricultural productivity that heavily disturbed the national economies of many countries and harsh their social stability. Locust outbreaks are usually transboundary swarm movements and often are international in nature. It increased the regional cooperation between affected countries for better management programs that are essential. Exclusive management tactics need to be used by continuous support both moral and financial at all national and international levels. This will ensure timely and successful modern/emerging technologies for sustainable preventive management of desert locusts.

Modern initiatives with comprehensive technologies provide matchless opportunities to overcome the problem. The combined use of modern technologies will strengthen and build the capacities of underdeveloped economies. This will allow prevention and control over drastic damage by desert locusts. These emerging technologies provide us with all locust breeding areas that are easily accessed without any security threat.

KEYWORDS

- **technologies**
- **monitoring**
- **forecasting**
- **desert locust**
- **global information system**

REFERENCES

Bais-Moleman, A. L.; Schulp, C. J. E.; Verburg, P. H. Assessing the Environmental Impacts of Production- and Consumption-Side Measures in Sustainable Agriculture Intensification in the European Union. *Geoderma* **2019,** *338*, 555–567.

Cigliano, M. M.; Torrusio, S. Geographic Information Systems (GIS) and Remote Sensing in Entomology: Studies in Grasshoppers and Locusts (Orthoptera: Acridoidea). *Revista de la Sociedad Entomológica Argentina*. **2017,** *62* (1–2).

Cressman, K. Desert Locust. In *Biological and Environmental Hazards, Risks, and Disasters*; Sivanpillai, R., Ed.; Elsevier: Amsterdam, The Netherlans, 2016; pp 87–105.

Cressman, K. Role of Remote Sensing in Desert Locust Early Warning. *J. Appl. Remote Sens.* **2013,** *7*, 075098.

Crook, D. R.; Robinson, B. E.; Li, P. The Impact of Snowstorms, Droughts and Locust Outbreaks on Livestock Production in Inner Mongolia: Anticipation and Adaptation to Environmental Shocks. *Ecol. Econ.* **2020,** *177*, 106761.

Deveson, E. D. Satellite Normalized Difference Vegetation Index Data Used in Managing Australian Plague Locusts. *J. Appl. Remote Sens.* **2013,** *7*, 075096.

Dewan, A. M.; Yamaguchi, Y.; Rahman, M. Z. Dynamics of Land Use/Cover Changes and the Analysis of Landscape Fragmentation in Dhaka Metropolitan, Bangladesh. *Geo J.* **2012,** *77*, 315–330.

FAO (Food and Agriculture Organization of United Nations) Madagascar Locus Crisis. Food and Agricultural Organization of the United Nations, 2018. http://www.fao.org/emergencies/crisis/madagascar-locust/en

Gijbels, M.; Marchal, E.; Verdonckt, T. W.; Bruyninckx, E.; Vanden Broeck, J. RNAi-Mediated Knockdown of Transcription Factor E93 in Nymphs of the Desert Locust (Schistocerca gregaria) Inhibits Adult Morphogenesis and Results in Supernumerary Juvenile Stages. *Inter. J. Mol. Sci.* **2020,** *21* (20), 7518.

Gómez, D.; Salvador, P.; Sanz, J.; Casanova, C.; Taratiel, D.; Casanova, J. L. Desert Locust Detection Using Earth Observation Satellite Data in Mauritania. *J. Arid. Envir.* **2019,** *164*, 29–37.

Han, X. J. *Study on Remote Sensing Mechanism and Methods for East Asian Migratory Locust Hazard Monitoring*; University of Chinese Academy of Sciences: Beijing, 2003 (in Chinese with English Abstract).

Herold, M.; Achard, F.; DeFries, R.; Mollicone, D. Earth Observation and Political Negotiations: Linking Requirements and Capabilities in the Context of the UNFCCC/REDD Process. In *Proceedings of the 33rd International Symposium on Remote Sensing of Environment (ISRSE)*, May 4–8; Stresa, Italy, 2009; pp 1–4.

Hüttich, C.; Gessner, U.; Herold, M.; Strohbach, B. J.; Schmidt, M.; Keil, M.; Dech, S. On the Suitability of MODIS Time Series Metrics to Map Vegetation Types in Dry Savanna Ecosystems: A Case Study in the Kalahari of NE Namibia. *Rem. Sens.* **2009,** *1*, 620–643.

Joga, M. R.; Zotti, M. J.; Smagghe, G.; Christiaens, O. RNAi Efficiency, Systemic Properties, and Novel Delivery Methods for Pest Insect Control: What We Know So Far. *Front. Physiol.* **2016,** *7*, 553.

Kanagasabai, L. Solving Optimal Reactive Power Problem by Shark Reek, Amplified Locust Search, and White Rhinoceros Algorithms. *Aust. J. Elect. Electron. Engin.* **2020,** *17* (3), 211–223.

Klein, I.; Oppelt, N.; Kuenzer, C. Application of Remote Sensing Data for Locust Research and Management—A Review. *Insects*. **2021,** *12* (3), 233.

Latchininsky, A. V.; Sword, G.; Sergeev, M.; Cigliano, M. M.; Lecoq, M. Locusts and Grasshoppers: Behavior, Ecology, and Biogeography. *Psyche* **2011,** *2011*, 578327.

Lawrence, P. L.; Chase, T. N. Investigating the Climate Impacts of Global Land Cover Change in the Community Climate System Model. *Int. J. Clim.* **2010,** *30*, 2066–2087.

Lazar, M.; Piou, C.; Doumandji-Mitiche, B.; Lecoq, M. Importance of Solitarious Desert Locust Population Dynamics: Lessons from Historical Survey Data in Algeria. Èntomol. *Exp. Et. Appl.* **2016,** *161*, 168–180.

Low, F.; Waldner, F.; Latchininsky, A.; Biradar, C.; Bolkart, M.; Colditz, R. Timely Monitoring of Asian Migratory Locust Habitats in the Amudarya Delta, Uzbekistan Using Time Series of Satellite Remote Sensing Vegetation Index. *J. Environ. Manag.* **2016**. http://dx.doi.org/10.1016/j.jenvman.2016.09.001.

Ma, J. W.; Han, H. Z.; Hasibagan, H. Z.; Wang, C. L .; Zhang, J. Y.; Zhi, Y. X.; Deveson, T. Monitoring East Asian Migratory Locust Plagues Using Remote Sensing Data and Field Investigations. *Int. J. Remote Sens.* **2007,** *26* (3), 629–634.

Mohammed, L.; Diongue, A.; Yang, J. T.; Bahia, D. M.; Michel, L. Location and Characterization of Breeding Sites of Solitary Desert Locust Using Satellite Images Landsat 7 ETM+ and Terra MODIS. *Adv. Entomol.* **2015,** *3*, 6–15. https://doi.org/10.4236/ae.2015.31002

Penner, M. P.; Simpson, S. J. Locust Phase Polyphenism: An Update. *Adv. Insect Physiol.* **2009,** *36*, 1–272.

Renier, C.; Waldner, F.; Jacques, D. C.; BabahEbbe, M. A.; Cressman, K.; Defourny, P. A Dynamic Vegetation Senescence Indicator for Near-Real-Time Desert Locust Habitat Monitoring with MODIS. *Remote Sens*. **2015,** *7*, 7545–7570.

Shen, N.; Shen, Z. A Study on Use of MODIS Data to Detect the Relationships Between Outbreak of Oriental Migratory Locust and Vegetation Growth: A Case Study of the Area Huanghua County, Hebei Province. 2005. Nanjing Normal University.

Showler, A. T.; Ould Babah Ebbe, M. A.; Lecoq, M.; Maeno, K. O. Early Intervention Against Desert Locusts: Current Proactive Approach and the Prospect of Sustainable Outbreak Prevention. *Agronomy.* **2021,** *11* (2), 312.

Silva, S.; Alçada-Almeida, L.; Dias, L. C. Development of a Web-Based Multicriteria Spatial Decision Support System for the Assessment of Environmental Sustainability of Dairy Farms. *Comput. Electron. Agric.* **2014,** *108*, 46–57.

Singh, N. S.; Sharma, R.; Parween, T.; Patanjali, P. K. Pesticide Contamination and Human Health Risk Factor. In *Modern Age Environmental Problems and Their Remediation*; Oves, M., Khan, M. Z., Ismail, I. M. I., Eds.; Springer, 2018; pp 49–68. https://doi.org/10.1007%2F978-3-319-64501-8_3

Sivanpillai, R.; Latchininsky, A. V. Mapping Locust Habitats in the Amudarya River Delta, Uzbekistan with Multi-Temporal MODIS Imagery. *Environ. Manag.* **2007,** *39*, 876–886.

Sivanpillai, R.; Latchininsky, A. V.; Driese, K. L.; Kambulin, V. E. Mapping Locust Habitats in River Ili Delta, Kazakhstan, Using Landsat Imagery. *Agric. Ecosyst. Environ.* **2006,** *117*, 128–134.

Sugahara, R.; Tanaka, S.; Jouraku, A.; Shiotsuki, T. Geographic Variation in RNAi Sensitivity in the Migratory Locust. *Gene.* **2017,** *605*, 5–11.

Tratalos, J. A.; Cheke, R. A. Can NDVI GAC Imagery Be Used to Monitor Desert Locust Breeding Areas? *J. Arid Environ.* **2006,** *64*, 342–356.

Veran, S.; Simpson, S. J.; Sword, G. A.; Deveson, E.; Piry, S.; Hines, J. E.; Berthier, K. Modeling Spatiotemporal Dynamics of Outbreaking Species: Influence of Environment and Migration in a Locust. *Ecology* **2015,** *96*, 737–774.

Whyard, S.; Singh, A. D.; Wong, S. Ingested Double-Stranded RNAs Can Act as Species-Specific Insecticides. *Insect Biochem. Mol. Bio.* **2009,** *39* (11), 824–832.

Wynant, N.; Santos, D.; Wielendaele, P.; Broeck, J. Scavenger Receptor-Mediated Endocytosis Facilitates RNA Interference in the Desert Locust, *Schistocerca gregaria. Insect Mol. Biol.* **2014,** *23*, 320–329.

Yang, P. Y.; Ren, B. Y. Promoting the Application of Green Crop Pest Management Technologies—Review on the Key Issues in the National Technical Schemes of Major Crop Pest Management for 2011 to 2017. *Plant Prot.* **2018,** *44*, 6–8 (in Chinese).

Yao, X.; Zhu, D.; Ye, S.; Zhang, N.; Li, L. Spatial Interpolation Methods Study Based on Geostatistics for the Grasshopper Population. *Sensor Lett.* **2014,** *12*, 645–650.

Yao, X. C.; Zhou, D. H.; Yun, W. Y.; Peng, F.; Li, L. A WebGIS-Based Decision Support System for Locust Prevention and Control in China. *Com. Electron. Agric.* **2017,** *140*, 148–158. https://doi.org/10.1016/j.compag.2017.06.001.

Zhang, L.; Lecoq, M.; Latchininsky, A.; Hunter, D. Locust and Grasshopper Management. *Annu. Rev. Entomol.* **2019,** *64*, 1.1–1.20.

Zotti, M. J.; Smagghe, G. RNAi Technology for Insect Management and Protection of Beneficial Insects from Diseases: Lessons, Challenges and Risk Assessments. *Neotrop. Entomol.* **2015,** *44*, 197–213. DOI: 10.1007/s13744-015-0291-8

CHAPTER 6

Locust Attacks on Crop Plants and Control Strategies to Minimize the Extent of the Problem

TABINDA ATHAR[1], HINA FATIMA[2], AISHA A. WARIS[1], AQSA[3], NAFISA KANWAL[3], and FARAH MAJID[4]

[1]*Institute of Soil and Environmental Sciences, University of Agriculture Faisalabad, Faisalabad, Pakistan*

[2]*School of Applied Biosciences, Kyungpook National University, Daegu, South Korea*

[3]*Department of Soil Science, University of Agriculture Faisalabad, Sub Campus Burewala-Vehari, Pakistan*

[4]*Department of Entomology, University of Agriculture, Faisalabad, Faisalabad, Pakistan*

ABSTRACT

Locusts' invasion has been a major threat to crop plants and overall food security across the globe. Crop damage by locusts is even more threatening and devastating due to global pandemics such as COVID-19. Recently, the locust's attack has caused major damage to the agricultural and food crops and the value of this damage is estimated to cross billions of dollars. Although these locusts and grasshoppers are major feeders of crop plants, at the same time their populations are also important to maintain ecosystem functioning and balance. However, their overpopulation in certain seasons

Locust Outbreaks: Management and the World Economy. Umair Riaz, Khalid Rehman Hakeem, (Eds.)

can cause major destruction of growing crops. Therefore, it is significantly important to check their populations, feeding activities, habitats, and crop damage in various regions so that effective management measures and strategies can be devised as per requirements and extent of the problem. The control and management measures mentioned in this chapter can greatly help the farmers and researchers to minimize the locust's invasion to achieve food security on a sustainable basis.

6.1 INTRODUCTION

Locusts belong to the superfamily *Acridoidea* and are orthopterous pests. They are significant from an economical point of view as they attack numerous crops and cause destructions all around the world. The recent locust attack was beginning at the time of crop harvesting. Twenty-two countries were affected by locust attacks. Food and Agriculture Organization evaluate that the locust invasion in Ethiopia caused the devastation of cropland area over 197,163 ha, cereal loss of 356,286 metric tons, and about 1 million people of Ethiopia alone were affected with starvation. COVID-19 epidemic itself has the aptitude to cause the crisis of food security in Africa. Regional agriculture foodstuff supply also faced disturbance, as well as transport, approach to inputs, labor movement, and roadblocks reasoned to COVID-19. Specifically, the supply of pesticides and further tools required to restrain the swarm of locusts were also affected by the pandemic. Collectively both disasters created the situation of starvation, poverty, and sickness. Another factor that contributes to the locust crowd was climate change. Additionally, unexpected windstorms and rainfall in the Arabian head extent of locusts increased due to more vegetative growth as they possessed a large area to feed on.

In July 2019 three severe attacks have been observed in North Gujarat. According to the Agriculture department of Gujarat 17,000 ha areas were infected with locust swarms since 1993–1994. That was the largest attack and huge losses of 5–6 Crore have been reported (Joshi et al., 2020). Locusts are the oldest known migratory insects and are very devastating due to their biology and ecology. Massive outbreak of locusts has threatened crop production, terrestrial environments, and food security all over the globe. Especially, the extent of the problem is being increased due to global warming and climate change and it caused egg spawning of 2–400,000 locust eggs per square meter.

In 2019–2020, in several districts of Gujarat and Rajasthan severe locust attack was observed. According to the Government of Rajasthan 179,584 ha area was affected by the attack. Gujarat Government has described that 19,313-ha area of two districts was affected and crop losses have been observed throughout 2019–2020. Locust outbreaks in Pakistan, India, and East Africa are significantly important for the migrations of more than 150 km per day and they consume more food equivalent to the weight of their bodies on daily basis.

Locust activity largely depends on temperature. Locust activity will be at a higher level when they maintained their optimum body temperature between 35 and 40°C. Locust can also damage crops at little temperature. But their action will be drop or stop below 15°C temperatures. According to an estimate, nymphs can consume 100–450 mg of vegetation each day and mature locusts can consume 0.2 g per day. The following figures are strongly variable depending upon different features such as densities, crowd size, type of crop, stages of the crop, and temperature (Hao et al., 2020). Therefore, strong research collaboration is required among scientific communities of different regions to minimize the damage and threats to global food security.

6.2 CROPS DAMAGED BY LOCUST ATTACK

Desert locusts consume wide varieties of crops including cereal crops, vegetable crops, fruit trees, and coffee (reviewed in COPR. Brytania, 1982). Locusts cause destructions of green plants mostly, but they also prefer to eat pleasant crops. Additionally, if green vegetation is present then locusts have less attraction toward mature crops. If the moisture level and quality of seed are low locusts are more likely to attack the seedling crops at the two to fourth leaf stage. However, oilseeds and winter pulses are less commonly damaged by the attack.

The extent of damage by locust attack on various crops is discussed below:

6.2.1 COTTON CROP

The cotton crop is attractive to damage. If the damage occurs at the early stage of flowering, the extent of loss on yield is high and will be less afterward. As the cotton flowering starts adult locusts also appear in the summer zone area which increases the danger of loss in these zones (Bullen, 1972).

6.2.2 WHEAT AND BARLEY CROPS

The basic crop can be severely damaged, specifically when they are at their harvesting stage in the spring breeding zone. Complete loss of grain is caused when locust eat the last moist portion of the plant (Bullen, 1972).

6.2.3 BULRUSH MILLET CROP

It is the major grain crop of Indo-Pakistan dessert and Sahara desert. It is the favorite grain crop of locusts. This crop is largely grown in that area where locusts mostly grow; they eat mature grains and leaves which caused huge damage (Bullen, 1972).

6.2.4 MAIZE CROP

The heavy damage due to locust occurs in several parts of Eastern Africa where maize is grown under desiccated and hot conditions. The damage is very severe as they eat up all leaves of the plant and also cobs are eaten away (Bullen, 1972).

6.2.5 SORGHUM CROP

Not all types of this crop are liked by locusts throughout their growing periods, but the mature grains often attract locusts. The newly formed crowd of locust cause major damage (Bullen, 1972).

6.2.6 SUGARCANE CROP

Damage to sugarcane crops due to locust depends upon the variety and growth stages. For instant, in Pakistan damage is high at the early stages of growth (Bullen, 1972).

6.2.7 PASTURE DAMAGE BY LOCUST ATTACK

Green pasture is favored by locust attack. Freshly sown grasses and shorter grasses are vulnerable to locust attack. Instead of feeding on forbs, Italian locust swarms are linked with grazing pasture and will assault on alfalfa and fodder grasses (COPR, 1982). According to a recent estimate, destruction was narrated to pasture with very little effect on other crops in Argentina. Nymphs and mature locusts were found out on various shrub and pastures in Paraguay (Medina et al., 2017).

If green pasture weeds and other crops are present in the area, then Lucerne crops are less disposed to attack. Recently sown Lucerne crop which has five leaves is more susceptible to damage. But if the Lucerne has a crown with five leaves, then it has a superior chance to recover from harm. While the Lucerne crop is green and retains this characteristic for a long time when some other crops are found they are in focus. If the crown is hardly damaged it means the plant is less susceptible to death.

6.2.8 FRUIT TREE DAMAGE BY LOCUST ATTACK

An immature swarm of locusts has desired to attack fruit trees as they need a settlement place for their growth. Date, orange, pawpaw, lemon, and grapevine are most susceptible to be damaged by the locust's attack. If the locust attacks only once in a year its effect remains more than a year after. For example, in Morocco orange trees were damaged by the attack in a period between 1954–1955 (Bullen, 1972).

6.2.9 CEREAL CROPS DAMAGE BY LOCUST

Cereals crops are an essential part of the human diet every day, provide energy and nutrient directly via human consumption and indirectly via meat because it comprises chief livestock feed. Locust plagues pose an emerging threat to global crops, causing laborious disruptions (Adel et al., 2021). Major cereal crops are susceptible to locust attacks that can severely affect their productivity. The intensity of locust damage is highly variable depending on seasonal conditions and population density. In the Middle East, locusts had been known as a crop pest, probably the time of first cultivation was being practiced almost 10,000 years ago. Locusts damage, even sporadic in both time and space, can destroy crops entirely and almost entirely over an enormous area, causing physiological depression to humans and local famine (Bullen, 1966). These four species of locusts that are responsible for locust plague in the region are red locusts (*Nomadacris septemfasciata*), brown locust (*Locustana pardalina*), Desert Locust (*Schistocerca gregaria*), and African migratory locust (*Locusta migratoria migratorioides*).

6.3 TYPES OF LOCUSTS CAUSING CROP DAMAGE

6.3.1 RED LOCUST

Its life cycle starts from October, the female locusts lay about three egg pods (100 egg present per pod) in moist soil. After 30 days of the incubation period, first instar hopper hatch and develop within 50–60 days. It produces only one generation per year. Generally, its coloration is brown, reddening, and deepening with age and it is distributed in most of the Cape Verde Islands, Africa south of Sahara, Mauritius, and Madagascar. In about the last 130 years, it was major species of three plagues. It was reported South Africa spent £933,000 on grassland and crop protection measures, during the previous major red locust plague from 1930 to 1944.

6.3.2 BROWN LOCUST

The female locusts lay about three to five egg pods (45 eggs present per pod) in moist soil. After 15 days of the incubation period, first instar hopper hatch and develop within 25–30 days. It produces three generations per year. Brown locust species has rarely green, brown, tegmina with scattered brown spots and is distributed in Angola, Zambia, Namibia, South Africa, Zimbabwe, Botswana, and Lesotho. The brown locust is known as a pest in Southern Africa. It was reported South Africa spends mean annually between £100,000 and £200,000 on control of brown locusts. In 1985–1986, about 250,000 hopper bands and 40,000 adult swarms were controlled.

6.3.3 DESERT LOCUST

The female locusts lay about two to four egg pods (60–160 egg present per pod) in moist soil. After 10–70 days of the incubation period, first instar hopper hatch and develop within 22–70 days. It produces three generations per year. Desert locust is identified because of larger size. Postrenal tubercle blunt, straight, and slightly inclined backward and distributed throughout the world, but they have been recorded in the East Atlantic Ocean, Mozambique St. Helena, Red Sea, British Island, Arabian Sea, the Mediterranean region, and the Persian Gulf. Desert locusts are a major pest, and four factors are responsible: (i) the movement of a swarm, (ii) the range of parts and plants eaten, (iii) individual intake of food, (iv) frequency of swarms. They eat fresh vegetation per day equal to their body mass and cause damage to the plant by cutting of uneaten leaves and stems and branches breaking due to densely settling of population.

6.3.4 AFRICAN MIGRATORY LOCUST

The female locusts lay about one to five egg pods (30–100 egg present per pod) in moist soil. After 30 days of the incubation period, first instar hopper hatch and develop within 24–35 days. It produces four to five generations per year. Except for prosternum tubercle absence and larger size, distinctive features are wings with light yellow tinting, branches end with a smoky appearance and black anal veins. Migratory locust is a widely distributed species in the world. Its population can be categorized into two different lineages Northern lineages which contain the Eurasian continent and Southern lineages which contain Africa, southern Asia. In the late plagues of 1941, African migratory locust swarms damage extensively cereal crops area of Africa south of the Sahara. At the end of the 1920s and the early 1930s, severe plagues of the desert locust and tree locust were there, out of which migratory locust was the most voracious crop pest. The lower-lying grass-growing areas in Southeast Asia are most commonly attacked by the oriental migratory locusts. The estimate and reports about the value of crops being destroyed are less, but this type of locusts prefer feeding on grass species.

Moreover, they can also feed on wheat, tobacco, sweet potato, sugarcane, soybean, sorghum, sisal, palm, sago, rice, pineapple, pea, millet, maize, lettuce, groundnut, coconut, citrus, beans, barley, banana, and bamboo. Also tree locus has been reported as a major locust pest in Senegal to Sudan on millet crops. Most of its attack in Sudan has been reported on ornamental trees, shade trees, dates, fruit trees, cotton, and *Acacia senegal*. But it is a minor pest in North Africa and the Middle

East and causes damage to trees, fruits, vegetables, grapevines, dates, and tobacco.

Javanese grasshopper is a major locust pest in West Malaysia and Java and causes significant damage to the trees and cash crops. Mostly, it damages the young plants, growing shoots, and leaves. The largest infestation of Javanese grasshopper was observed in the Teak forests in 1915. Generally, the plants that are attacked by the Bombay locusts are also prone to be attacked by Javanese grasshopper.

6.3.5 THE ITALIAN LOCUSTS

This tree-feeding locust belongs to the subfamily *Calliptaminae* and is described as a pest. Crops that are important economically such as sunflower, wheat, bean, and many others are under attack by Italian locusts. They live on the ground surface and are often found on forbs and grass (Batáry et al., 2007). Italian locusts will attack mostly on alfalfa, grazing field, forage such as fodder. When locust attack on cropland from minor grasses, economical losses caused. These locusts are also considered a pest for the wheat crop (Sergeev and Van'kova, 2008).

6.3.6 BROWN LOCUST

Brown locusts belong to *locustana pardalina* family and cause huge damage to pasture and cereals in South Africa (Price and Brown, 2000). Changes in patterns of rainfall have been linked with extension and reduction. When rainfall increased in the summer season grasses become denser and taller which caused a reduction in locust range (Kieser et al., 2002). They have a great influence on animal pastures. Though efforts depend on the legal needs of growers to convey attacks on their assets, then insecticides are used to kill immature and mature swarms (Crooks and Cheke, 2014).

6.3.7 SAHELIAN TREE LOCUST

This tree pest belongs to the subfamily *Cyrtacanthacridinae*. The young nymph is responsible for damage on grasses and mature will eat crops including maize, millets, rice, and cotton. Gum Arabic is under attack, which is the major crop produced in the region of Sudan and has significant value for semi-nomadic natives (Haroon et al., 2011). The Sahelian tree locust is happy to live in an arid environment. They can resorb liquid from frass in dehydrated conditions. It does not compete with grass grazer because of tree feeding nature, but it is an important competitor for camel and cattle when they feed on trees (COPR, 1982). In addition, these locusts take part in the recycling of nutrients in Silvopastural classification. According to a study Nitrogen deposits were found in the area of Oman during locust attacks (Robinson, 2001).

6.3.8 DESSERT LOCUST

Dessert locusts feed on a huge amount of green vegetation such as fodder, crops, pasture and it is the most devastating and mobile pest in the world. These locusts are mostly found in the area of South Africa and the phase changing process is rarely found in this type (Chapuis et al., 2017). The last huge outbreak caused 80–100% losses of the crop in affected regions of Saharan Africa; it occurred in 2004–2005 (Brader et al., 2006). Desert locusts live in the area that receives an annual rainfall of 0–400 mm such as Africa and Northwest India. Outbreaks arise when they receive rainfall sufficient to increase locust growth (Cressman, 2016). The locust causes huge losses of fodder crops which is responsible for the decrease in livestock production. As pesticides had adverse effects on health an alternative would be used; according to that farmer will submit annual fee which would be later used in case of the outbreak (Abou-Ali and Belhaj, 2008).

6.3.9 CENTRAL AMERICAN LOCUST

Central American locusts, known as *Schistocerca piceifrons piceifrons*, caused huge agricultural damage in Mexico and Costa Rica state of America (Cullen et al., 2017). Intensive damage on beans, peanut, soya, cocoa, cotton sugar in areas of Central America is treated by using pesticides every year (SENASICA-D GSV, 2019). They are generally herbivores and found in regions of America. For the previous 30 years, the forest is converted into grassland due to ranching in the state of Yucatan has been observed. Forty-seven percent area is converted into grassland from 1981–2014 which provides a favorable region for the growth of *S. piceiforns* and is also responsible for swarming (Poot-Pech et al., 2016).

6.3.10 SPUR-THROATED LOCUST

Spur-throated locust also known as *Austracis guttulosa* is mostly found in Australian insects and belongs to Acrididiae family. It feeds on a huge number of crops including sorghum, soybean, wheat, sunflowers, eucalyptus, citrus; also it consumed several types of grass and native trees but it has less economic impact than other Australian pest. But this species can cause damage when the threshold level reaches 20 nymphs per m^2 (DPI, 2018).

This Brown locust adapted to an environment of wet summer and dry winter. Single swarm is grazing region varying in size from 200 to 800 ha which wholly defoliated trees as well as damaged pastures (Casimir and Edge, 1979). Nymphs of Spur-throated locust often move from rangeland to crops; then effective management for this purpose is adapted (DPI, 2018).

6.4 BOMBAY LOCUST

Bombay locust is also known as *patanga succincta.* It is normally a solitary pest and shows swarming behavior mostly in India. In Thailand Bombay, locust is the most famous food insect used for human consumption and the main agricultural pest (Phiriyangkul et al., 2015). It feeds on huge varieties of crops including banana, maize, nut, rice, millet, cabbage, citrus, castor, cucumber, groundnut, mango, guava, cashew, oil palm, rubber, sweet potato, and tea (Roffey, 1979). When the field converted from forest to grassland its population becomes increased. The area of the lalang grass field carries the locust densities from 20 to 50 per m^2.

6.5 RED LOCUST

The binomial name of the red locust is *Nomadacris septemfasciata* and distributed in Sub-Saharan Africa. Red locust has been restrained by International Red Locust Control Organization, since the previous attack that affects the African countries (Bahana, 2000). It feeds mostly on grasses but also on trees and shrubs. They can also compete with animals for feed. Though in the latest year, the red locust attack is minimum and does not damage the crop especially after the use of biopesticides (FAO, News Article: Red Locust disaster in Eastern Africa prevented, 2009).

6.6 AUSTRALIAN PLAGUE LOCUST

It is also named *Chortoicetes terminifera* belonging to the family Acrididae and found mostly in Australian insects. It is now the most important agricultural pest in Australia (Hunter, 2004). One of the most recent outbreaks was observed in 2010, with about 2,000 land farmers reporting problems (Deveson, 2011). It is mostly found in arid and semiarid regions with some variable precipitations (Hunter et al., 2001). Australian Plague locusts are mostly present in the area of animal grazing due to grasses. But the increase in grassland due to deforestation causes the expansion of this locust in that region (Deveson, 2011, 2012).

6.7 MIGRATORY LOCUST

Migratory locust belongs to a family *Locusta migratoria*, divided into two main subspecies based on the geographic range which are *L. migratoria migratorioides* and *L. migratoria migratoria*. In the whole world, this locust is a widely spread grasshopper species (Ma et al., 2012). It is present around the tropical and temperate hemispheres. It is the most significant pest in Uzbekistan, Russia, and Kazakhstan (Latchininsky, 2010). Due to human activities of agriculture, the population of migratory locust increased in Sudan, Mali, and Chad. For example, Sahara Dessert was inappropriate for this locust, but after human activities of agriculture its population increased. Migratory locusts have mostly damaged cereal crops, studies showed that first instars attack on pasture grass, and later the third instars fed on winter wheat (Qing and Chunxian, 2008).

6.8 SIBERIAN LOCUST

Siberian locust is also known as *Gomphocerus sibiricus* which belongs to the subfamily Gomphocerinae and is the most significant agricultural pest in some eastern areas of Europe including the steppe zone of Siberia.

Also, it is an important pest in Xinjiang. It severely causes damage to oat, wheat, barley, corn, tobacco, cabbage, mustard, and also pasture. But, the Siberian locust does not affect wild grazing in Switzerland (Spalinger et al., 2012).

6.9 DAMAGE CAUSED BY LOCUSTS IN DIFFERENT SEASONS

6.9.1 DAMAGE TO AUTUMN CEREALS CROPS

Green cereal crops are more attractive to locusts, the reason is the growing stages of crops. Firstly, locusts like to graze on short green forage and secondly, primary cereals crops are crop up as swarms from delayed summer hatching. Autumn 2010 experience suggested, prefer feed is being wheat crops tracked by barley and triticale. It seems that oats crops are usually only eaten when the choice is inadequate or further vegetation is drying off.

6.9.2 DAMAGE TO SPRING CEREALS CROPS

Damage to green cereal crops consists of node chewing lower the head which leads it toward fall off and chewing of bracts and awns that surround the developing grain. But it is uncommon when locusts chewed the milky-dough stage of grain. The remaining moist plant parts are eaten by locusts which causes complete loss of grains.

6.9.3 DAMAGE TO SUMMER CEREALS CROPS

Locust attack is being vulnerable to rice seedling. Drill sown crops are being more affected than aerial sown crops as locusts will be inhibited by water. If growing edges have eaten, the plant growth will be affected. At later stages of growth, swarms damaged the cereals to the waterline. To save the further plant, water should hold a bit deeper. At the vegetative stage, rice can tolerate realistic damage. Sorghum crop is less susceptible to locust damage; it has robust seedling that emerges sturdily and grows quickly.

6.9.4 TYPE OF LOCUST DAMAGE TO CEREALS CROP AND EFFECT ON YIELD

The most common type of damage is being defoliation which ultimately affects the crop yield. Besides leaf destruction, it damages the crop plants in different ways as locusts swarm eat maize's cobs and male inflorescences too, while in wheat, stems inferior to ripen ears may be chewed, and grain, both unripe and ripe, is eaten. Serious pests of cereal crops are migratory locusts of Africa. In 1931, they destroyed 40% and 50% respectively of the maize and wheat crops of Kenya during their peak infestation year.

It was investigated locust attack decreases the maize yield to zero at the time of 3–5 in. high or at flowering stage. However, it turns down the maize yield to only 20% whether it attacked after the grain set yet it was quite unripe (White 1946). Locust's attack is being restricted by some cereals varieties as sorghum. A possible reason for the susceptibility variation may be hydrocyanic acid which is present in most sorghums crops and its amount differs with the growth stage and the variety (Snowden 1936). The resistant varieties have greater acid content in young plants but reduce with maturity, mostly at the time of the flowering stage onward. The quantity of acid is high in plants that are affected by extreme weather. Analyses show, sorghum crops are more susceptible to damage at the time of harvesting because of low acidity (Dalziel 1937).

6.10 MANAGEMENT TECHNIQUES TO CONSTRAIN LOCUST SWARMS

Locusts are known as major agriculture pests worldwide. Locust swarms have been causing pandemonium for centuries because of their serious infestation in the Indian subcontinent and the Middle East. Controlling numerous swarms that are causing a massive loss in agriculture and food security of many countries requires international involvement over the only government. There is a need to introduce the concrete solution in this field; however, throughout this COVID-19 pandemic, locust swarms would be painless to constraint via adapting advanced comprehension of pest ecology and biology along with controlling techniques and developing monitoring (Ghosh et al., 2020). There are distinct techniques to control locusts swarms viz locating suitable region and managing locust plague.

- LOCATE LOCUST SUITABLE REGION

Gigantic locust swarms can be inhibited via the identification of the regions that possess suitable condition to locust emergence as a droughty region that recently receives heavy rainfall. Then any small-sized swarms can be tracked down, initiating the monitoring program. The tracking action can be done by local volunteers, drones, and satellites that can aid the process. Swarms can be eliminated if detected.

- MANAGE LOCUST PLAGUE

After the locust outbreak, formers should use all the weapons that are available to them. The most probably weapons are some methods. That description is given below.

6.11 TRADITIONAL METHOD

Mostly, only farmers are important to do anything while facing locust attacks. They alternatively built up certain cultural and physical control before the chemical utilization. Particularly opposed crops small plant infections, these methods are carried out in a group or alone with chemical control. For example, various farmers conjoin the pesticide's application with fire which aids them to keep away the locusts in the evening time.

6.11.1 SPRINKLING OF FLOUR

Locusts and grasshoppers can be malnourished with flour treatment which gums their mouth. Applying a thin layer of flour will be more efficient to the crops.

6.11.2 USING OF MEDICATED NETS

There is an increasing reliance on anti-insect nets being a vital IPM (Integrated pest management) tool for their act as the best greenhouse and protection for the plant and crops. It also minimizes the demand for pesticide application via blocking the ingress of insects and pests into the crop environment. If it is sprayed with neem and garlic oil which is highly efficient for the crop and protects it from the insects like grasshoppers and locusts.

6.11.3 EMPLOYING OF NEEM OIL

There is an advanced methodology of utilizing neem oil with Azadirachtin to keep insects from growing into their fleeting swarms. It prevents the hormones and the pheromones formation expected to keep pace with the yellow-and-black convivial structure, which outbreaks scorch the Middle East and Africa. When neem oil is applied to insect abdomen, it devastates their antennae (Rangiah and Gowda, 2019). Odisha government advises the farmers to spray 200 liters of 5% neem seed kernel extract (NSKE) solution per acre of land during the afternoon to inhibit insect attacks. They can also spray a solution by adding 300 PPM neem insecticide in 200 L of water.

6.11.4 APPROACHING OF GARLIC OIL

Garlic oil spray is cost-effective, homemade, and more advantageous for crops as it is a chemical-free and organic product which odor can aid to deter a few garden pests and grasshoppers. Making a spray is the best way to apply the mix to the plants without damaging the vegetable plants or flower buds. These organic sprays will stick at a dry, cool, dark place for up to 2 weeks and are most helpful in reining the pest.

6.11.5 SCIENTIFIC METHOD

Locusts are certain species of grasshoppers. Due to changes in density locusts are capable to render their appearance and behavior. In fewer numbers, they avoid each other, and colors were nondescript but in the case of huge numbers, they create swarm and brightly colored. Therefore, it can be used as a sound basis to better conditions for controlling locusts' populations.

6.11.6 GENETICALLY MODIFIED METHOD

The use of variant bug sprays, pheromone, bug-safe cultivars improvement, insect attractant traps, organic control by predators as well as a parasite, sterile males emancipation, and growth regulators are the module in the management strategy of bug sprays and pesticides. Growth regulators are less worthwhile as insecticide; however, integrated pest management can be effectively used in locusts controlling through easily manageable strategies (Courtright and Kumaran, 1986). One of the most eco-friendly and safe techniques is the utilization of sterile males and the sterility of males within pervasive populations originated either through cytoplasmic or chromosomal incompatibility.

6.12 USE OF VARIOUS CHEMICALS

6.12.1 FENITROTHION (C9H12NO5PS)

It is used for locust control on pasture, some crops, and a wide range of cereals, for example, soybean, grazing lucerne, sorghum, grapes, apples, lettuce, tomatoes. It is activated by ingestion and contact with insects. Some ovicidal activity is done via this chemical because it affects the nervous system of the insect which leads to rapid insect death.

6.12.2 MALATHION

Malathion is a pesticide that is utilized to kill insects on stored products, crops, golf courses, outdoor sites in-home gardens where shrubs, trees are grown at home, and also used to kill medflies and mosquitoes over large outdoor areas. Normally, it is sprayed on crops and over a large area by an airplane over wide land areas. It usually breaks down via water, sunlight, and soil bacteria within a few weeks but retains for several months in the environment (Moharana et al., 2020).

6.12.3 BIOPESTICIDES

Some of the fungi sold as Green Guard like *Metarhizium acridum* family fungi are a biological control agent that can be used in sensitive areas. These agents are derived from a naturally occurring *Metarhizium anisopliae* (Australian fungus) which attacks locusts. This Green Guard can be used to treat only immature locusts because it takes 8–18 days to affect the lifecycle. Generally, locusts are contaminated with the attachment of Metarhizium spores. After germination, these spores penetrate the internal organs and are developing the cuticle of the insect. After the treatment, mortality happens within 8–21 but it can take time in the cool temperature. It is advisable to treat the crops or plants when locusts are in nymphal stages which are less nomadic than the adult ones (Shuvo et al., 2020).

6.12.4 BOTANICAL PESTICIDES

The recent research work on botanical pesticides reveals a promising result. A study demonstrated orange peel, wintergreen, and caraway oils are effective against locust when used with a linseed oil/bicarbonate emulsion. This formulation caused 100% mortality of desert and migratory locusts within 24 h but not more field experiments are conducted (Abdelatti and Hartbauer, 2020).

6.13 PREVENTIVE MEASURES

- At night farmers should monitor the field in the grouping because millions of insects set in the field between seven and nine to rest.
- Dig large ditches around the field, play loud instruments, and excavate ditches that covered the entire field and apply insecticide dust in them.
- Locusts congregate on shrubs and trees in the evening.
- Spray 1500 ppm per liter of Azadirachtin that is a neem-based insecticide on crops as a preventive measure.
- In the early morning and late night, spraying should be done. At these times, locusts can be killed effectively because they gather in huge numbers on the bushes. Spraying gives a lot of insect control.

Some bug insecticides sprays utilized in control programs of desert insect have progressed over the constant use of Dieldrin, organo-chlorines, Aldrin, BHC, and Lindane to organophosphates. Nowadays, the most common insecticides in India for controlling desert insects are malathion and fenitrothion. In terms of activity, organophosphates are contact bug sprays (Roffey et al., 1970).

6.14 RECENT ADVANCES IN LOCUST CONTROL

The fresher innovation in controlling insects came from 1993. From the west, enormous scope swarm attacks this year were constrained via cleaning BHC on container group and settled multitudes in which splash bug fluid sprays through the air before it spread over a large area. The BHC utilization was restricted in 1993 because it uses a lot of shower fluid litter volumes known as ULV (ultra-low volume) splashing; it was used for the first time during the 1950s against the Desert Locust. It is normally utilized and the most productive strategy as it applies between 0.5 and five shower fluid liters per hectare. Although, somewhere the range of 0.5 and 1.0 liter/hectare is noted favorable as ULV beetle control.

Nowadays, invasions of desert insects have showered with ULV (Ultra Low Volume) definitions of contact pesticides through the use of Micro-nair AU 8115, ULVAmast, and microULVA sprayer on the ground. The ULV splash strategy is intended to cover all areas of little beads through shower onto grasshoppers at low portion rates. Because of hostile change, the desert insect control has been viable to locusts' strategies that are increasingly improved wellbeing for administrators, animals, occupants, and nature.

6.15 FUTURE PROSPECTS

Schistocerca gregaria is the most destructive locust in desert conditions and is reported to cause most losses all over the globe. The changing weather influenced by the global warming and climate change phenom-enon created the conditions that specifically favored the rapid reproduc-tion and migration of these locusts. Therefore, a significant spreading was observed in West Africa, West Asia, Arabian Peninsula, East Africa, and many other countries in the world. The damage caused to the crops has been so severe that more than 25 million people in Tanzania, Uganda, South Sodan, Somalia, Kenya, and Ethiopia will be suffering from food-insecure conditions.

The use of advanced technology such as drone systems can offer significant ease for the management measures for locust control. There is an exigent urgent need to develop standard operating procedures for the optimal and rationalized use of technology. Drones can easily help to manage the aerial and ground management measure for the control of locusts in huge areas. Drone technology can offer effective spraying on the targeted areas and thereby causes a significant reduction in infestations that otherwise would have been impossible by using manual spraying methods. Especially, the smaller swarms can be easily targeted that may escape aircraft control and aerial sprays in the inhabited agricultural lands or areas. Latest drone technologies are versatile and can be adapted for carrying chemical payloads. The use of spectral cameras and ultra-low volume spraying technologies offers targeted spray in specific areas. Moreover, drones also offer mapping facilities to determine crop damage and losses.

Moreover, the causes behind the rapid reproduction and spreading of all locust types are unknown and therefore their overall management and control are difficult. Their vast growth, reproduction, and migration are directly associated with the favorable climatic conditions and availability of food sources. Usually, they are more attracted to the warmer but high rainfall areas, so effective management measures should be specially optimized for these areas. The probability of locust migration and their reproduction in these areas may be enhanced in near future due to the climate change phenomenon, and therefore policies and control measures must be regulated and implemented according to prevailing conditions in specific areas.

Early identification of their egg-laying sites is also important to reduce the magnitude of the problem and locust population. Continued forecasting and monitoring of locusts' population in the specific sites should be prioritized for optimized control. There should be the proper establishment of locust controlling units and must be monitored by the appropriate ministries. Safe and wise use of pesticides is also a good approach for timely control and management of locusts' populations. The use of efficient and environmentally friendly products is even more important to avoid killing nontargeted insects. Moreover, there should be proper training of spray teams to handle and manage the pesticide application and storage.

People in the invaded areas should be trained and alerted to keep the livestock and crops safe by efficiently spraying the targeted areas only. Metarhizium is often recommended to control locusts as it is developed by using skin fungus of locusts by the international institute of tropical agriculture. Furthermore, the use of certain chemicals may cause environmental concerns, especially when these are applied by the means of aerial sprays. The fallen droplets of these pesticides can cause contamination of water sources and soil and thus may cause significant damage to the growing plants and crops. Besides they may also cause some burns, skin rashes, eye problems, neurological and respiratory disorders.

There is a concept of manufacturing and bringing larger quantities of certain pesticides to the invaded sides. Many countries and regions suffer serious dilemmas for purchasing pesticides by proper planning according to the extent of infestations. Generally, the prediction of locust outbreaks for longer terms is not an easy approach as it has the periodic nature of the outbreak. All of these pesticides are not being used for locust control, and therefore the management and proper storage of leftover pesticides is the biggest challenge. Improper handling and storage can cause significant problems for soil, the environment, animals, beneficial insects, and human beings. Smaller amounts of pesticides and synthetic chemicals can be purchased and can be sent to the affected areas by air cargo on short notice. It will offer rapid response to the specific areas and the creation of larger stocks will also be avoided. Besides detailed assessments should also be conducted before ordering the new stocks.

The analysis and statistics have shown that more than 100 million human beings are prone to be threatened by the locust attack. Therefore, it is significantly important to use technological approaches for locust monitoring, controlling their population, and predicting outbreaks. Remote sensing has become an important technological option for data sourcing and is being used since 1980 for locust management activities and improved control measures. Full capacities of available remote sensing techniques and applications have not been fully exploited. Locust management strategies are complex and therefore multidisciplinary targeted approaches are required including the good collaboration of researchers from biology, ecology, and entomology background. Moreover, studying the behavior of organisms, weather predictions, interaction with multiple species of organisms, climate analysis, and spatial distribution modeling is significantly essential to ensure effective control. Optimized use of bio-agents and remote sensing applications is even more effective for valuable locust management. Food and Agriculture Organization (FAO) has introduced standard monitoring procedures for data collection.

GIS techniques can help to map and monitor the habitat of locusts in different environmental conditions that can promote the transitional processes between gregarious and solitarious phases. Moreover, hatching times, and outbreak timings can also be predicted according to weather forecasting, current monitoring of vegetation, and historical information.

The use of artificial intelligence-based systems and drone technologies offers significant potential to improve locust control by capturing their real-time pictures with the help of higher resolution cameras. The produced image is processed by using MATLAB and is then used for recognizing swarms and insects by Max pooling and Convolutional neural network algorithm methods. This system is only being used in few countries or by few research and development institutes. Its use should be properly promoted to all countries for efficient and effective locust control. This type of advanced technology can greatly improve the food security and socioeconomic conditions of nations and societies. Moreover, locusts are being used as food sources by different nations, so proper detection and identification of swarms can also be used to catch the locusts. Therefore, locusts can be caught by using advanced techniques and later be sold or used as a food source.

Furthermore, fungal pathogens can also be used for the effective control of locusts and grasshoppers. These fungal pathogens can even be directly sprayed to the locusts. Moreover, parasites can also be used to attack the locusts and therefore excellent control of the locust population can be made possible. Even if the caught locusts are not being used as food by certain communities, they can be utilized to make high-quality organic fertilizers by the composting process. Compost is a continuous and slow source of nutrients and helps to improve soil quality on a sustainable basis along with the significant cutting off the use of inorganic fertilizers.

Furthermore, barrier treatments should also be encouraged to prevent the locust's attack. Barrier treatment is focused on using parallel vegetation strips. These vegetation strips are properly treated with good quality and recommended pesticides, but the areas between these vegetation strips are not sprayed. Usually, the locusts and hoppers move in the downward direction for searching food, so they will encounter the sprayed vegetation strips and thus a lethal dose of pesticides will be accumulated in their bodies. There are numerous advantages of barrier treatment as it allows efficient control of larger areas and thus is an ideal solution for survey and control-related activities. One of the best advantages of this management strategy is that a major portion of the vegetation area remains untreated from these pesticides and thus nontargeting killing of certain organisms is also reduced. Furthermore, these barriers offer sustainable management due to the minimum associated cost.

6.16 CONCLUSION

Locusts have caused significant destruction all over the globe and therefore identification of different species, their life cycle, crop damage, and habitats are important for effective management and control. There are many techniques and approaches to control locust invasion, but all of them cannot be used in all areas due to certain variations in prevailing environmental conditions and feeding requirements. But all pesticides and control measures may not be safe for all the areas and therefore there should be a good evaluation of the efficacy and quality of the treatments. So, the identification of the right methods and technologies is essentially important to ensure the elimination of locusts. If their populations are left unchecked for certain reasons, they can grow exponentially and can produce millions of offspring in the shortest duration. Locust attacks will be continually causing food security threatening in different countries as they are rapidly migrating to the desired locations. So collective efforts of researchers, communities, organizations, scientists, students, and governments are required to combat this problem on a sustainable basis.

KEYWORDS

- **climate change**
- **desert locust**
- **red locust**
- **cotton**
- **sustainability**
- **management tools**

REFERENCES

Abdelatti, Z. A. S.; Hartbauer, M. Plant Oil Mixtures as a Novel Botanical Pesticide to Control Gregarious Locusts. *J. Pest Sci* **2020,** *93* (1), 341–353.

Abou-Ali, H.; Belhaj, M. *Cost Benefit Analysis of Desert Locusts' Control: A Multicountry Perspective.* Paper presented at the Economic Research Forum Working Papers (Giza), 2008.

Bahana, J. *The Role of the International Red Locust Control Organisation for Central and Southern Africa (IRLCOCSA) in the Management of Migratory Pests.* Paper Presented at the Workshop on Research Priorities for Migrant Pests of Agriculture in Southern Africa, Plant Protection Research Institute, Pretoria, South Africa, March 1999; 2000.

Batáry, P.; Orci, K. M.; Báldi, A.; Kleijn, D.; Kisbenedek, T.; Erdős, S. Effects of Local and Landscape Scale and Cattle Grazing Intensity on Orthoptera Assemblages of the Hungarian Great Plain. *Basic Appl. Ecol.* **2007,** *8* (3), 280–290.

Brader, L.; Djibo, H.; Faye, F.; Ghaout, S.; Lazar, M.; Luzietoso, P.; Babah, M. O. Towards a More Effective Response to Desert Locusts and Their Impacts on Food Security, Livelihoods and Poverty. *Multilateral Evaluation of the 2003–05 Desert Locust Campaign*; Food and Agriculture Organisation: Rome, 2006.

Brytania, W. *The Locust and Grasshopper Agricultural Manual*: Centre for Overseas Pest Research, 1982.

Bullen, F. Locusts and Grasshoppers as Pests of Crops and Pasture-a Preliminary Economic Approach. *J. Appl. Ecol.* **1966,** 147–168.

Bullen, F. *Review of the Assessment of Crop Losses Caused by Locusts and Grasshoppers.* Paper presented at the International Study Conference on the Current and Future Problems of Acridology, London, 1970. Proceedings, 1972.

Casimir, M.; Edge, V. The Development and Impact of a Control Campaign Against Austracris guttulosa in New South Wales. *Pans* **1979,** *25* (3), 223–236.

Chapuis, M. P.; Foucart, A.; Plantamp, C.; Blondin, L.; Leménager, N.; Benoit, L. et al. Genetic and Morphological Variation in Non-Polyphenic Southern African Populations of the Desert Locust. *Afr. Entomol.* **2017,** *25* (1), 13–23.

Cheke, R. A.; Rosenberg, L.; Kieser, M. *Workshop on Research Priorities for Migrant Pests of Agriculture in Southern Africa, Plant Protection Research Institute, Pretoria, South Africa, 24–26 March 1999*; Natural Resources Institute, 2000.

Courtright, J. B.; Kumaran, A. K. A Genetic Engineering Methodology for Insect Pest Control: Female Sterilizing Genes. In *Biotechnology for Solving Agricultural Problems*; Springer, 1986; pp 325–336.

Cressman, K. Desert Locust. *Biological and Environmental Hazards, Risks, and Disasters* **2016,** 87–105.

Crooks, W. T.; Cheke, R. A. Soil Moisture Assessments for Brown Locust Locustana Pardalina Breeding Potential Using Synthetic Aperture Radar. *J. Appl. Remote Sens.* **2014,** *8* (1), 084898.

Cullen, D. A.; Cease, A. J.; Latchininsky, A. V.; Ayali, A.; Berry, K.; Buhl, J. et al. From Molecules to Management: Mechanisms and Consequences of Locust Phase Polyphenism. *Adv. Insect Physiol.* **2017,** *53*, 167–285.

Dalziel, J. M. *The Useful Plants of West Tropical Africa*, 1937.

Deveson, E. The Search for a Solution to Australian Locust Outbreaks: How Developments in Ecology and Government Responses Influenced Scientific Research. *Historical Rec. Aust. Sci.* **2011,** *22* (1), 1–31.

Hao, Z.; Drake, V. A.; Taylor, J. R.; Warrant, E. Insect Target Classes Discerned from Entomological Radar Data. *Remote Sens.* **2020,** *12* (4), 673.

Haroon, W. M.; Pages, C.; Vassal, J.-M.; Abdalla, A. M.; Luong-Skovmand, M.-H.; Lecoq, M. Laboratory and Field Investigation of a Mixture of Metarhizium acridum and Neem

Seed Oil Against the Tree Locust Anacridium Melanorhodon Melanorhodon (Orthoptera: Acrididae). *Biocontrol Sci. Technol.* **2011,** *21* (3), 353–366.
Hunter, D. M. Advances in the Control of Locusts (Orthoptera: Acrididae) in Eastern Australia: From Crop Protection to Preventive Control. *Aust. J. Entomol.* **2004,** *43* (3), 293–303.
Hunter, D.; Milner, R.; Spurgin, P. Aerial Treatment of the Australian Plague Locust, Chortoicetes terminifera (Orthoptera: Acrididae) with Metarhizium, 2001.
Joshi, M. J.; Raj, V.; Solanki, C. B.; Vaishali, V. Desert Locust (Schistocera gregaria F.) Outbreak in Gujarat (India). *Agriculture and Food: E-Newsletter* **2020,** *2* (6), 691–693.
Kieser, M.; Thackrah, A.; Rosenberg, J. Changes in the Outbreak Region of the Brown Locust in southern Africa, 2002.
Latchininsky, A. Locusts, 2010.
Le Gall, M.; Overson, R.; Cease, A. A Global Review on Locusts (Orthoptera: Acrididae) and Their Interactions with Livestock Grazing Practices. *Front. Ecol. Evol.* **2019,** *7*, 263.
Ma, C.; Yang, P.; Jiang, F.; Chapuis, M. P.; Shali, Y.; Sword, G. A.; Kang, L. Mitochondrial Genomes Reveal the Global Phylogeography and Dispersal Routes of the Migratory Locust. *Mol. Ecol.* **2012,** *21* (17), 4344–4358.
Medina, H. E.; Cease, A.; Trumper, E. The Resurgence of the South American Locust (Schistocerca cancellata). *Metaleptea* **2017,** *37* (3), 17–21.
Moharana, S.; Khuntia, P.; Das, S. S.; Das, R.; Panigrahi, S. A Systematic Review of Behavioral Aspects & Management Techniques to Control Locusts. *Int. J. Sci. Adv. Res. Technol.* **2020,** *6* (7), 361–369.
Phiriyangkul, P.; Srinroch, C.; Srisomsap, C.; Chokchaichamnankit, D.; Punyarit, P. Effect of Food Thermal Processing on Allergenicity Proteins in Bombay Locust (Patanga succincta). *Int. J. Food Eng.* **2015,** *1* (1), 23–28.
Poot-Pech, M.; Ruiz-Sánchez, E.; Ballina-Gómez, H.; Gamboa-Angulo, M.; Reyes-Ramírez, A. Olfactory Response and Host Plant Feeding of the Central American Locust Schistocerca piceifrons Walker to Common Plants in a Gregarious Zone. *Neotrop. Entomol.* **2016,** *45* (4), 382–388.
Qing, Y.; Chunxian, Y. Feeding Habits and Economic Threshold of Locusta migratoria tibetensis. *Acta Phytophylac. Sin* **2008,** *5*.
Rangiah, K.; Gowda, M. Method to Quantify Plant Secondary Metabolites: Quantification of Neem Metabolites from Leaf, Bark, and Seed Extracts as an Example. *The Neem Genome*; Springer, 2019; pp 21–30.
Robinson, M. D. Desert Nitrogen Cycles: Fecal Nitrogen from a Population of the Sahelian Tree Locust in Oman. *Sci. Technol.* **2001,** *6*, 33–38.
Roffey, J. Locusts and Grasshoppers of Economic Importance in Thailand. *Locusts and Grasshoppers of Economic Importance in Thailand* 1979 (14).
Roffey, J.; Popov, G.; Hemming, C. Outbreaks and Recession Populations of the Desert Locust Schistocerca gregaria (Forsk.). *Bull. Entomol. Res.* **1970,** *59* (4), 675–680.
Sergeev, M.; Van'kova, I. Dynamics of the Italian Locust Calliptamus Italicus L. Population in the Southeast of the West Siberian Plain. *Contemp. Prob. Ecol.* **2008,** *1* (2), 204–209.
Sharma, A. Locust Control Management: Moving from Traditional to New Technologies–An Empirical Analysis. *Entomol. Ornithol. Herpetol.* **2014,** *4* (141), 2161–0983.1000141.
Shuvo, S. B.; Tusher, M. A.; Ghosh, M. A Technical Approach to Prevent Locust Swarm, 2020.

Snowden, J. D. *The Cultivated Races of Sorghum*, 1936.

Spalinger, L. C.; Haynes, A. G.; Schütz, M.; Risch, A. C. Impact of Wild Ungulate Grazing on Orthoptera Abundance and Diversity in Subalpine Grasslands. *Insect Conserv. Divers* **2012,** *5* (6), 444–452. https://doi.org/10.1111/j.1752-4598.2011.00180.x

Van Huis, A.; Cressman, K.; Magor, J. I. Preventing Desert Locust Plagues: Optimizing Management Interventions. *Entomologia Experimentalis et Applicata* **2007,** *122* (3), 191–214. https://doi.org/10.1111/j.1570-7458.2006.00517.x

White, R. M. Preliminary Observations on Some Effects of Artificial Defoliation of Wheat Plants. *Sci. Agric.* **1946,** *26* (5), 225–229. DOI: 10.4141/sa-1946-0026

Xu, Z.; Elomri, A.; El Omri, A.; Kerbache, L.; Liu, H. The Compounded Effects of COVID-19 Pandemic and Desert Locust Outbreak on Food Security and Food Supply Chain. *Sustainability* **2021,** *13* (3), 1063.

CHAPTER 7

Effect of Entomopathogenic Fungi on Migratory Locusts

MIRZA ABDUL QAYYUM*, SHAFQAT SAEED, NAEEM IQBAL, ARSLAN KHAN, UNSAR NAEEM-ULLAH, HASAN RIAZ, MUHAMMAD ISHTIAQ, MUHAMMAD FIAZ, MUHAMMAD UMAIR SIAL, ABOUBAKAR SIDDIQUE, and MUHAMMAD AQEED MEHDI

Institute of Plant Protection, MNS University of Agriculture, Multan, Pakistan

ABSTRACT

The locust outbreaks are known to occur across different regions around the globe and have a significant impact on global food security. Large swarms of locust proved havoc to many agricultural croplands and pasturelands in a very few time-period. The swarm of locusts in the recent past represents one of the worst infestations over the past many years in many countries including Ethiopia, Somalia, India, Kenya, Pakistan, Oman, Saudi Arabia, and Tanzania. The locusts were striving for every little piece of green mass on the earth during swarming and thriving potential crops like cotton, wheat, maize, and vegetable to risk. Toxic insecticides proved negative as hazardous effects on human health, biodiversity, and ecosystem services are bearing some serious abnormalities. The number of chemical insecticides proved costly as millions of dollars were spent during the control program. The use of green

Locust Outbreaks: Management and the World Economy. Umair Riaz, Khalid Rehman Hakeem, (Eds.)

technologies proved effective for sustainable crop pest management programs around the world as an alternative to synthetic pesticides. So, biological pest management is best suited as an alternative to synthetic insecticide using natural microorganisms like fungi potentially known as entomopathogens. In the recent past, approximately 150 isolates belong to *Beauveria* and *Metarhizium* species against desert locusts out of which 50 isolates proved lethal. Among *Metarhizium, Metarhizium falvoviride* was first reported as the potential bio-control agent. Till now many species including *Beauveria bassiana, Metarhizium anisopliae,* and *Isaria fumosorosea* proved lethal, but *Metarhizium anisopliae var. acridum* proved best and environmentally much safer to use and have no significant effects on natural bio-control agents. However, the availability of substrates at the local level, the abundance of indigenous virulent fungal isolates, and cheap labor are necessary requirements for the establishment of a successful mycotoxin-based industry.

7.1 INTRODUCTION

The desert locust, *Schistocerca gregaria* (Forskål) (Orthoptera), undergoes three immature stages viz egg, nymph (hopper), and adult. Locusts belong to the grasshopper family *Acrididae* that is native to multiple scrub habitats including African deserts, Sahel (African regions), and Arabian desert Peninsula including Saudi Arabia, Oman, and Yemen. It has also caused a plague in Southwest Asian countries like Pakistan, Iran, India, and Afghanistan (Showler, 2002). Desert locusts differ from grasshoppers because their life cycle is multiphasic: solitarious and gregarious. In the former state, population density is very low and there is no economic threat. When conditions become favorable for breeding followed by droughts around the coastal areas of the Red sea, Indo/Pak border, Mauritania, and Sahel Mountainous regions, an increase in population density takes place which results in competition for food ultimately inducing a transition phase in locusts. It triggers a rapid change in physiology, behavior, and morphology (changes in color and shape) of locusts on a regional scale (Uvarov, 1921; Roffey and Popov, 1968). Resultantly, they enter into the transient phase hence called transient, and cluster in densely populated bands of wingless nymphs (hoppers) and swarms of winged adults. The number of hoppers is correlated with their behavior. This behavioral change takes place during basking, roosting feeding, and sheltering in vegetation. The transient phase is very crucial regarding locust future attack and pandemic because it is an intermediate condition that may run either toward gregarization or dissociation. Gregarization results in the expansion of these insects from southern Australia (Tanzania) to Guinea, southern Europe, and Bangladesh (Cressman, 1996). Band's density depends on the behavior of hoppers and the age of nymphal instars, which also vary according to the habitat and weather conditions. In some studies, more than 30,000 hoppers of first instar are recorded/$m^2$and just 1000 per m^2 in case of fifth nymphal instar.

Locust is a migratory pest that's why chemical control is not the only weapon to control this pest. Besides, chemical control poses a serious threat to the environment and is also very expensive in locust control (Brader, 1988; Everts, 1990). In the plague of 1988, a plethora of insecticides (13M L) was sprayed in more than 10 countries of northern and northwestern Africa at the cost of USD 100 million (Anonymous, 1990). For a couple of years, desert areas of developing and under developed countries are the habitats of locusts; that is why it impossible for them to combat this potential pest unaided. There is an imminent need to provide an economical and environmentally friendly substitute in these locust-affected areas. IPM is a multifactorial strategy that has a diversity of domains, helpful to control locusts. Biological control is one of the safest methods in pest control operations. Biocontrol agents should be exploited and used against pests. Efforts should be made to focus on the theory of "Diseases against Diseases." In this domain, the discovery of Entomopathogens has revolutionized the field of insect pest control. Entomopathogenic Fungi, especially *Beauveria bassiana* and *Metarhizium anisopliae* from order *Hypocreales*, are highly host-specific and eco-friendly bio-pesticides (Lacey and Goettel, 1995; Gillespie et al., 2000).

Their local production is feasible and can be easily propagated. These Entomopathogenic fungi kill their host (insect) by direct action through penetrating its cuticle (Welling et al., 1995; Webb et al., 1998).

7.2 HISTORICAL PERSPECTIVE OF USING ENTOMOPATHOGENIC FUNGI

Brazil and China have been successfully managing insect pests of agriculture and forestry by using these biocontrol agents since the 1970s. A decline in their production and commercialization was seen for two consecutive decades because of the negligence of government and industry. Statistical reports mention that during 2008, more than a million-hectare area under sugarcane cultivation was treated with *Metarhizium anisopliae* to control spittlebugs (Li et al., 2010). Recent statistics told us that almost 170 fungal-based insecticides and acaricides have been successfully registered and commercialized in the last three decades (Faria and Wraight, 2007). Countries having governmental support for initial viable inoculum, mass production of EPF, and successful field applications for the first successful microbial pest-control program have multiple success stories. However, the availability of substrates at the local level, the abundance of indigenous virulent fungal isolates, and cheap labor are necessary requirements for the establishment of a successful mycotoxin-based industry.

7.3 TOXICITY OF ENTOMOPATHOGENS AGAINST LOCUSTS

Locusts have been practically prohibited by the use of Entomopathogenic fungus *Beauveria bassiana* on a large scale (Mason and Erlandson, 1994). These fungi have special secretions that are toxic metabolites and are produced either on artificial media or in field conditions (Mazet et al., 1994, Vey, 1998, Vey et al., 2001). Most of these mycotoxins have antibacterial and insecticidal action, due to the presence of cyclic peptides they have low molecular weight and are secondary metabolites (Strasser et al., 2000, Vey et al., 2001). *Hirsutella thompsonii* is a fungal pathogen that produces an insecticidal protein hirsutellin A (HtA), the only macromolecular compound purified to homogeneity to date (Mazet and Vey, 1995; Liu et al., 1996). The reason why these Entomopathogens are still lagging behind in pest control is their slow rate of action because infection starts when fungal spores come in contact with insect cuticle, they require at least 2–3 days to infest or kill their host, researchers are keen to improve the commercial efficacy and viability of this potential bio-agent with optimized culture production, ecological stability, and proper application methods of the viable inoculums (St Leger and Screen, 2001). Similarly, a spore-producing microsporidium, *Paranosema (Nosema) locustae*, has been considered a potential pathogen for controlling desert locusts; followed by the ingestion it directly infects the midgut tissues of orthopteran, weakens, and eventually kills the host (Herrero-Galán et al., 2013).

7.4 EFFICACY OF ENTOMOPATHOGENIC FUNGI UNDER LAB CONDITIONS AND IN FIELD

Mycotoxins have considerable potential to control acridids (Goettel et al., 1995). Fungal formulations based on oils of *Metarhizium anisopliae var acridum* (Driver et al., 1999) showed positive results against locusts in the southern and western deserts of Africa (Scherer et al., 1992; Douro-Kpindou et al., 1995). Technology advancement proved to be very promising in locust control, especially Ultra-low Volume (ULV) application techniques and multiple surfactants, adjuvants, adhesives combined with active ingredients (mycotoxins) enhanced virulence of conidial spores (Symmon, 1992; Bateman, 1997). LUBILOSA, a French-based biological locust combat program in Africa (Lomer et al., 2001; Langewald et al., 1997), Commonwealth Scientific and Industrial Research Organization Program in Australia (Hooper et al., 1995) conducted field trials and proved the Entomopathogenic behavior of *Metarhizium anisopliae var acridum* was proved effective as significant lethality was recorded to locusts and grasshoppers. Quesada and Vey (2003) proved the toxicity of 90/2-Dm isolate of *B. bassiana* on the nymphal instars of *Locusta migratoria*. Beauvericin and Bassianolide are two low molecular weight proteinic compounds, biologically active against locusts. Another fungal isolate EABb 90/2-Dm having high molecular weight by *B. bassiana* is specific in its mode of action against Moroccan locust, *Dociostaurus maroccanus* (Orthoptera: Acrididae) especially at the molecular level (in vitro). In Africa, under laboratory conditions, insecticidal efficacy of fungal strain CQMa102 isolated from *Metarhizium anisopliae var. acridum* (formerly *Metarhizium flavoviride*) (Ascomycota: Hypocreales) (Driver et al., 2000; Hooper et al., 1995) has been evaluated against desert locust, *Schistocerca gregaria* (Wang et al., 2003).

7.5 APPLICATION TECHNIQUES OF ENTOMOPATHOGENS

Biopesticides, an ecofriendly alternative way of integrated pest management, have target-specific nature and pose the least negative effects on the agroecosystems. These biopesticides differ based on biologically active ingredients they have such as bacteria, viruses, fungi, nematodes, or naturally occurring substances like plant extracts. Application techniques of these highly desirable crop protectants vary according to the nature of microorganisms, and formulation development is not an easy task. Commercial product formulation is a multifactorial process because it should be easily storable, economical, persistent, and convenient in handling also less degradable. The efficacy of any biocontrol agent depends upon its droplet size (if liquid) and deposition on plant parts. Biopesticide formulations are classified into liquid and dry formulations. Liquid formulations may be in multiple combinations depending upon carrier, that is, water, oil, and polymer-based. Among these, water-based formulations are successful in action when fortified with surfactants, adjuvants, adhesives, stickers, stabilizers, antifreezers, and additional nutrients. Spray drying, freeze-drying, or air drying are commonly employed dry formulation which either aided with or without fluidized bed (Tadros, 2005; Brar et al., 2006; Knowles, 2008). Some important biopesticide formulations along with their pros and cons are given below.

7.6 DUSTS

Particle size ranges from 50 to 100 μm. Active ingredient (10%) is adsorbed on finely ground powdery substrates like clay or talc. An application can be done manually or mechanically. Adsorption of dust can be enhanced by adding anticaking agents, adhesives, and ultra-violet protectants. Because of having adverse effects on users and serious inhalation hazards of specks of dust. they are now restricted (Knowles, 2001).

7.7 POWDERS FOR SEED TREATMENT

The active ingredient is mixed with a powdery material that acts as a carrier combined with some adherent for successful persistence. Sometimes, a colored pigment is added to the mixture to identify the treated seeds, mostly red color is used (Woods, 2003).

7.8 WATER DISPERSIBLE GRANULES

Like wettable powders, a uniform suspension is made by breaking up granules in a liquid carrier combined with wetting/dispersing agents. These formulations are advantageous because of having no inhalation problem as they are dust-free and have good storability (Knowles, 2008).

7.9 ENVIRONMENTAL FACTORS INFLUENCING FIELD EFFICACY

Application of fungus spores to host or foliage, spores decline by several factors such as sunlight, rain, humidity, foliage chemistry, and host behavior. Insects cuticle is invaded by fungal spores usually within 24 h. Germination within the cuticle is affected by several environmental factors.

Sunlight is a major component that effects field efficacy of fungal spores, in epigeal habitat UV-A and UV-B have mortality factor of fungal isolates on the insecticides containing phylloplane (Behle, 2006). In another word, fungal conidia has half-life under natural conditions, in sunlight, in viability percentage per unit area of 3–4 h (Braga et al., 2001).

Fungal propagules can be catastrophic by the application due to rain events. Several formulation emulsifiable suspensions and wet-able powder washed off leaves applied in water only. Temperature affects fungal insecticide's field efficacy. Growth rates of entomopathogenic insecticide range between 23°C and 28°C, growth rate slows down above 30°C and stops at 37. Similarly, germination of conidia is adversely affected by temperatures above 30°C.

7.10 TEMPERATURE AND THERMOREGULATION

In the case of field applications, *Metarhizium* is potentially slow and has variable killing speed on the different host which can impose practical issues for its application. Many grasshoppers and locusts are actively thermoregulate which sustains the temperature of the body at a set point for the day. Locusts and grasshoppers use the orientation of sun and wind, body gestures on vegetation to control their body temperature (Wellington et al., 1998). The temperature of the internal body can be considerably varying from outer temperature and maintained for a significant period in the given right environmental conditions. Species of locusts and some of the grasshoppers preferred body temperature in the range 38–40°C which is the upper limit for the germination of fungal spores.

Locusts thermoregulate to a body temperature can also have behavioral fever response for the growth of fungal infection. In this condition, the incubation time will be increasing due to thermoregulation of the body temperature optimally. Pathogens are unable to develop colony and the incubation period will be amplified accordingly. Killing speed can also be minimized due to night temperature due to drop of ambient temperature dramatically in desert conditions. Drop in temperature is also a favorable opportunity for pathogen over one day and provide long survival time in cool night-time conditions with optimum thermoregulatory circumstances during day time. But in cloudy, high temperature, optimum thermoregulatory circumstances lead to a low chance of pathogenicity.

Thermal ecology among pathogen and host is a vital part to determine the efficacy performance of given myco-insecticide.

7.11 OTHER ENVIRONMENTAL FACTORS

Primary factors are discussed which are environmental temperature, humidity, rain, and sunlight, but other factors are soil-related which are related and make the environment for interaction of host-pathogen such as soil texture, habitat pH, CE capacity of the soil, and salts. Soil structure, texture, and moisture are involved in a complex interaction that has involvement in insect movement. Soil pore space distribution influences the infectivity of fungal spores by physical contact.

Agricultural contributions like fertilizer, pH regulators, chemical pesticides, and other practices cause a major impact on microbial populations, but in the case of entomopathogens have been few studies found in a relationship with soil and pathogen titers rate.

Table of entomopathogenic products available:

Name	Trade name	Infected insect species
Beauveria bassiana Balsamo	Biorin, Bio-power, Daman, Racer, Naturalis, Nagestra,	Several arthropod pests including insects and mites
Beauveria brongniartii Vuill.	Bas-Eco	Lepidoptera and Coleoptera larvae
Hirsutella thompsonii Fish.	No-Mite	Mite species
Isaria fumosorosea Wize.	No-fly	Whitefly
Metarhizium anisopliae Sorokin.	Biomet/Ankush, Bio-Magic, Novacrid, Pacer	Several insects
Metarhizium brunneum	Attracap	Agriotesspp
Verticillium lecanii Zim.	Bio-Catch, Bioline/Verti-Star, Mealikil.	Bugs and soft-bodied insects including mealybugs and aphids
Lecanicillium lecanii Zim.	Vertalec, Lecatech-WP, Mycotal, Varunastra	Aphids, leafminers, mealybugs, scale insects, thrips, whiteflies

KEYWORDS

- **locust outbreaks**
- **food security**
- **green technologies**
- **entomopathogens**
- ***Metarhizium anisopliae* var. acridum**
- **climate change**

REFERENCES

Anonymous. A Plague of Locusts: Special Report. US Congress, Office of Technology Assessment. USA: US Government Printing Office, Washington DC, 1990.

Behle, R. W. Importance of Direct Spray and Spray Residue Contact for Infection of Trichoplusia in Larvae by Field Applications of *Beauveria bassiana. J. Econ. Entomol.* **2006,** *99* (4), 1120–1128.

Boyetchko, S.; Pedersen, E.; Punja, Z.; Reddy, M. Formulation of Biopesticides. In *Biopesticides: Use and Delivery Methods in Biotechnology*; Hall, F. R., Menn, J. J., Eds.; Humana Press: Totowa, NJ, 1998; pp 487–508.

Brader, E. W. *U.S. Patent No. 4,750,493*; Washington, DC: US. Patent and Trademark Office, 1988.

Braga, G. U. L.; Flint, S. D.; Miller, C. D.; Anderson, A. J.; Roberts, D. W. Both Solar UVA and UVB Radiation Impair Conidial Culturability and Delay Germination in the Entomopathogenic Fungus *Metarhiziumanisopliae. Photochem. Photobiol.* **2001,** *74* (5), 734–739.

Brar, S. K.; Verma, M.; Tyagi, R. D.; Valero, J. R. Recent Advances in Downstream Processing and Formulations of Bacillus thuringiensis Based Biopesticides. *Process Biochem.* **2006,** *41* (2), 323–342.

Cressman, K. Current Methods of Desert Locust Forecasting at Fao 1. *EPPO Bull.* **1996,** *26* (3–4), 577–585.

de Faria, M. R.; Wraight, S. P. Mycoinsecticides and Mycoacaricides: A Comprehensive List with Worldwide Coverage and International Classification of Formulation Types. *Biolog. Con.* **2007,** *43* (3), 237–256.

Everts, J. W. Environmental Effects of Chemical Locust and Grasshopper Control, 1990.

Gillespie, J. P.; Burnett, C.; Charnley, A. K. The Immune Response of the Desert Locust Schistocerca Gregaria During Mycosis of the Entomopathogenic Fungus, Metarhizium Anisopliae var Acridum. *J. Insect. Phys.* **2000,** *46* (4), 429–437.

Herrero-Galán, E.; García-Ortega, L.; Olombrada, M.; Lacadena, J.; Del Pozo, Á. M.; Gavilanes, J. G.; Oñaderra, M. Hirsutellin A: A Paradigmatic Example of the Insecticidal Function of Fungal Ribotoxins. *Insects* **2013,** *4* (3), 339–356.

Knowles, A. Recent Developments of Safer Formulations of Agrochemicals. *Environmentalist* **2008,** *28* (1), 35–44. DOI: 10.1007/s10669-007-9045-4

Knowles, A. Trends in Pesticide Formulations. *Agrow Reports*; PJB Publications Ltd. D215, 2001; pp 89–92.

Lacey, L. A.; Goettel, M. S. Current Developments in Microbial Control of Insect Pests and Prospects for the Early 21st Century. *Entomophaga* **1995,** *40* (1), 3–27.

Li, Z., Alves, S. B.; Roberts, D. W.; Fan, M.; Delalibera Jr, I.; Tang, J.: Rangel, D. E. Biological Control of Insects in Brazil and China: History, Current Programs and Reasons for Their Successes Using Entomopathogenic Fungi. *Biocont. Sci. Tech.* **2010,** *20* (2), 117–136.

Lomer, C. J.; Bateman, R. P.; Johnson, D. L.; Langewald, J.; Thomas, M. Biological Control of Locusts and Grasshoppers. *Ann. Rev. Entom.* **2001,** *46* (1), 667–702.

Mason, P. G.; Erlandson, M. A. Invitation Paper (CP Alexander Fund): The Potential of Biological Control for Management of Grasshoppers (Orthoptera: acrididae) in Canada. *Canadian Entomol.* **1994,** *126* (6), 1459–1491.

Roffey, J.; Popov, G. Environmental and Behavioural Processes in a Desert Locust Outbreak. *Nature* **1968,** *219* (5153), 446–450.

Showler, A. T. A Summary of Control Strategies for the Desert Locust, Schistocerca gregaria (Forskål). *Agric. Ecosyst. Environ.* **2002,** *90* (1) 97–103. ISSN 0167-8809. https://doi.org/10.1016/S0167-8809(01)00167-0.

Tadros, F. Applied Surfactants, Principles and Applications; Wiley-VCH Verlag GmbH and Co. KGaA, 2005; pp 187–256.

Uvarov, B. P. A Revision of the Genus Locusta, L.(= Pachytylus, Fieb.), with a New Theory as to the Periodicity and Migrations of Locusts. *Bull. Entomol. Res.* **1921,** *12* (2), 135–163.

Webb, S. C.; Hedges, R. E.; Simpson, S. J.; Diet quality influences the (&dgr;) 13C and (&dgr;) 15N of locusts and their biochemical components. *J .Exp. Biol.* **1998,** *201* (20), 2903–2911.

Welling, M.; Zelazny, B.; Scherer, R.; Zimmermann, G. First Record of the Entomopathogenic Fungus Sorosporella sp. (Deuteromycotina: Hyphomycetes) in Locusta migratoria (Orthoptera: Acrididae) from Madagascar: Symptoms of Infection, Morphology and Infectivity. *Biocontrol Sci. Tech.* **1995,** *5* (4), 465–474.

Wellington, W. G.; Johnson, D. L.; Lactin, D. J. Weather and Insects. In *Ecological Entomology*; Gutierrez, A., Ed.; Wiley & Sons: New York, 1998; pp 313–53; 850 pp.

Woods, T. S. Pesticide Formulations. In *AG*

CHAPTER 8

Locust Outbreaks, Climate Change, Sustainable Agriculture, and Environmental Effects

AYESHA HASSAN[1] and MUSTANSAR ASLAM[2]

[1]Department of Environmental Sciences, Faculty of Natural Sciences, University of Okara, Okara, Pakistan

[2]Department of Agricultural Sciences, Institute of Environmental & Agricultural Science, University of Okara, Okara, Pakistan

ABSTRACT

Locusts are swarming pests and they have started behaving gregariously in recent decades which turns into swarms and plagues and affects people, economy, and climate. Various species of locusts particularly desert locust (*Schistocerca gregaria*) are involved in causing outbreaks in regions of Africa and Asia (including India and Pakistan) and other continents which threatens multiple crops as well as human and agriculture well-being. Locusts infestations can be treated and controlled through careful planning and management at political and administrative levels and by educating farmers and local food producers. Global warming is considered the major cause of locust outbreaks in present years as warm climates support egg production and growth of locusts. The biological and ecological control is needed to combat locust outbreaks.

Locust Outbreaks: Management and the World Economy. Umair Riaz, Khalid Rehman Hakeem, (Eds.)

8.1 INRODUCTION

Locusts (Orthoptera: Acrididae) are a severe hazard to the environment and agriculture all over the world. Their outbursts can be found on each continent excluding Antarctica, putting the lives of nearly 10% of the global population in jeopardy. Locust species are well known due to their polyphagy, fervency, and the ability to migrate large distances. The bio-ecology of locusts has been elucidated through years of investigation. They occur in two forms, or "stages," that are interchangeable and density-dependent. Despite advancements in recognizing locust dynamics, human anticipation to forecast and regulate locust plagues remains elusive, as indicated by the fact that locust outbreaks continue to occur in the twenty-first century. The biggest reason for this is because locusts prefer to live in isolated, sparsely populated locations, and their geographical distribution often covers continents which makes monitoring and controlling locust populations extremely difficult.

Outbreaks can harm wildlife, farm animals, and farming, posing a threat to the local food production. Some grasshopper species go through a phenotypic transformation known as phase change when they are exposed to suitable circumstances, such as high grasshopper population growth. Grasshoppers transition from timid and solitary species to gregarious and migrating individuals capable of forming protracted migratory clusters during phase change. Locusts are classified as grasshoppers that have the capacity to go through this phase transformation (Pener and Simpson 2009). Although locust species exist in various subfamilies, their habit of flocking and dispersing is believed to have developed spontaneously multiple times within the grasshopper family (i.e., Acrididae). There are over 12,000 kinds of grasshoppers on the planet, but just 20 have exhibited the ability to convert into locusts (Song, 2010).

Multitude of insatiable locusts congregate into deadly swarms spanning thousands acres of land based on climatic conditions and vegetation cover (Topaz et al., 2012). The quantity of research publications on pests typically locust outbreaks and pests reflects their long-standing issue, which has sparked interest in biology, biochemistry, ecology, and evolutionary studies. The majority of this data is on locust infestation and its physiological characteristics, implying that the species' growth in population is exponential, resulting in disastrous environmental consequences like entirely consumed crop fields, creating food shortages and starvation, along with financial loss (Crook et al., 2020).

Locusts and grasshoppers have caused chaos on humans and animals since the dawn of time, and they continue to be a global concern today. The *Schistocerca gregaria* generally known as "The Dessert Locust" is still regarded as the most dreaded out of all locust species (Nevo, 1996). Uvarov in 1966 first described the theory of phase polymorphism in locust species namely *Locusta migratoria* (L.) which was then applied to other species as well (Pener and Yerushalmi, 1998). According to National Geographic, a swarm traveled from West Africa to the Caribbean in 1988, covering over 3100 miles (5000 km) in just 10 days.

Globally, weather conditions and precipitation pattern are the key elements that affect the diversity of any organism in space (Birch, 1957). Human population have faced locusts' outbreaks since primeval times (Anstey et al, 2009). Major aspects such as food, predators, parasites, and ailments are all mainly affected by the prevailing environmental conditions, which consequently influence the level of competition for essential resources. Climate and weather alterations affect the population's growth rate and genomic characteristics (Huffaker et al., 1984). Native life forms of plants and animal species have substantial amount of genetic diversity and variability and it is clearly evident that this genomic variation is sustained via natural selection as a result of organisms' adaptability to specific habitat (Huntly, 1991).

Conventional ground locust surveys are insufficient to handle the locust problem's large geographic extent in the restricted timeframe mandated by the pest's proliferation. In locust monitoring, remote sensing (satellite data) promises to be a potential approach. The desert locust and the Australian plague locust are the two locust species that are increasingly being monitored and forecasted using satellite data. Yet, several locust species have still to benefit from this geospatial technology (Akaninyene, 2020). Ever since the onset of 2020, the earth has been aware of a desert locust invasion stretching from Uganda to Pakistan. The states most directly impacted by such locusts include Kenya, Yemen, Uganda, Ethiopia, Somalia, Iran, and Pakistan (Dupe, 2020).

Since the dawn of agriculture, locusts have been humanity's worst foes. They can be found in numerous holy teachings like the Torah, Bible, and Quran. In the former book, they are known as Egypt's infamous Eighth Plague. There are around 100 allusions related to insects and other arthropods in the Old Testament, with the 40 citations to locusts and grasshoppers greatly surpassing all other related phrases. Flocks of locusts have often wreaked havoc on entire countries, causing famine. An ancient Roman narrator Pliny described the locust epidemic as a major cause of famine in the Roman regions of Numidia and Cyrenaica (now known as Algeria, Libya, and Tunisia) about 125 BC that took lives of 800,000 individuals. In Ethiopia, locusts destroyed 167,000 tons of food in 1958 that was sufficient to overcome hunger of one million people for a whole year.

Plagues caused by the desert locust species are primarily found in the deserts and scrublands of northern Africa, southwest Asia, and middle East. Swarms (extroverted adults that fly in larger groups) and bands (gregarious sprites that cannot fly but travel as a whole on the ground) of locusts are observed on an interregional scale during plagues and come from a variety of spawning regions as a part of an interconnected locust reproduction and migration pattern (Showler, 1995). Organisms with less-dense populations act like grasshoppers and are classified as solitarians though rarely in the severe form. Locusts may form the gregarious state and transform into long-term assemblies of nymphs (hopper bands) or adults (swarms), especially when they live in close connection (Hemming et al., 1979).

Locusts phase polymorphism is an unceasing and complicated yet density-dependent progression. However, species, strain within a species, sex, and pupal stage have an impact on phase traits and the extent of change. The solitary phase and extreme gregarious phase of locust phase polymorphism are ongoing processes which involves multiple variants. Crowding causes the gregarious state, as shown in Figure 8.1, which is known for its inclination to gather and build enormous swarms, whereas alienation causes the solitary state, in which individuals actively ignore fellow locusts (Uvarov, 1966). Moreover, environmental factors that are not density dependent can generally impact phase properties (Ariel and Ayali, 2015). However, phase features and the magnitude of change are affected by species, strain within a species. As a result, attempting to describe a whole and "ultimate" phase stage dependent on one or more phase attributes like color, appearance, morphometric, and behavior, etc. is fruitless (Pener and Yerushalmi, 1998).

FIGURE 8.1 Adult migratory locusts swarming in the air.

Source: Photo credit: CSIRO. https://creativecommons.org/licenses/by/3.0/

8.2 LOCUSTS OUTBREAK—EFFECTS ON AGRICULTURE

Locusts have been a serious global threat to agriculture and food security from many decades. A severe locust invasion in West Africa that occurred between 2003 and 2005 wreaked havoc on agriculture, killing $2.5 billion worth of crops which were intended for livelihood and trade purposes. Although $400 million in control measures, loss rates in certain areas topped 50% (Gall et al., 2019). Cease et al. (2015) claimed that human choices and land use practices can strongly impact locust population dynamics. They discovered methods that indicate how deteriorated grasslands alter the nutrient content of foliage in ways that enhance Mongolian locust outbreaks (*Oedaleus asiaticus*).

Locusts can be found on every continent except Antarctica. It ranges from distant regions of Africa to Asia. The desert locust has long been a major plant pest in several of Africa's poorest countries (Steedman, 1990). The capability of locusts to adjust to variations in population density through behavioral, physiological, and morphological way is one of their most remarkable biological traits. Severe locust outbreaks are threatening the physical environment and agricultural output in more than 100 countries worldwide, with Ethiopia, Somalia, and Kenya being the worst hit (Peng et al., 2020).

Certain pests (including locusts and grasshoppers) can develop into migrating locusts, which travel larger areas to connect ecosystem, populations, and land management practices. Grasshoppers are regarded as the prime factors residing in grassland ecosystems all over the world. They perform a critical part in trophic interactions and biogeochemical cycles. While many grasshopper species might pose a local risk to agriculture, locusts are of critical significance as they have the ability to jeopardize crop production and food security across huge areas. Migrating pests provide exceptional worldwide issues, necessitating coordinated multinational pest control measures (Toleubayev et al., 2007). Human influence on locust population demographics and distribution is well understood; however, human beings also impact locust population trends and dissemination patterns indirectly through land use and/or land forest cover alterations, which has not yet been studied and reported (Gall et al., 2019). The influence of the Moroccan locust, *Dociostaurus maroccanus*, seems to be reducing as well, most likely due to agricultural replacement (Latchininsky, 2013).

Locusts are discussed as a pest with exceptionally deleterious abilities: an adult desert locust can devour roughly its own muscle mass which is about 2 g/day in terms of fresh food. For instance, the dreaded locust attack happened in 1915 in UAE wiped out a majorly undervalued 536,000 tons of grains. As per the report of United Nation's Food and Agriculture Organization (FAO) in modern times, such as during the 2003–2004 locust infestations, it also equated to around US$30 million expended in anti-locust initiatives and awareness campaigns by the classic Africa (Ariel and Ayali, 2015).

Farm owners have also tried to safeguard fields against locusts. When locusts appeared in farmlands, however, they all have reached alarming levels and local farmers were unable to cope with pests. Farmers then turn to prayer or politicians for help (Lockwood, 2004). Plague locusts emerge from pandemic sites, which are natural environments where they grow and congregate. They are highly able to form swarms and travel to agricultural fields under extreme gregarious state, when they can cause serious harm. A preventive control strategy (PCS) purposes to control locust growth in the regulated, often remote yet not properly monitored outbreak areas (Van Huis et al. 2007). Though, regulating and managing locusts in such areas is clearly afar the capacity of individual farmers.

Popov (1987) claims that generally, the distribution and abundance of plague-causing locusts, primarily the Asiatic Migratory Locust and Italian Locust, are influenced by cyclical climate oscillations and that the eruption times of both species correlate. The researcher also believes that the size of epidemics is determined by farming techniques. By 1925, his research on population demographics of locusts in the Soviet Union has revealed a sequence of periodical rises and reductions in locust abundance. Because locust plagues are becoming more common, it is now critical to dramatically alter our strategy to altering, identifying, and addressing the problem, as well as to offer new and novel methods to locust challenges (Lecoq, 2005).

The desert locust feeds on a broad variety of vegetables and grain crops, as well as citrus, banana, groundnuts, fruit-producing trees, tea, as well as a diversity of other plants. This species is frequently regarded as one of the most harmful migratory pests in the whole globe due to its widespread distribution and importance to farming (Steedman, 1990). *S. gregaria*, known as Solitarious desert locust, avoid plant species carrying the alkaloid hyoscyamine, whereas gregarious locusts would aggressively feed on it, most likely to gain gut-content mediated toxicity and avoid predation. Furthermore, when given with fake meals with nutritional alterations in the lab, gregarious locusts are less discriminating than solitarious locusts (Simpson et al., 2002). Desert locust is extremely polyphagous species. It may target over 500 different plant species, including all farming systems. Perhaps a small 10 km^2 swarm can destroy 1000 ton of fresh flora per day, as 30 swarms could have 50 million individuals per square kilometers. Desert locust swarms are capable of invading a region of 29 million square kilometers.

Subsistence agriculture has a beneficial impact on locust populations in several cases, which could cause food insecurity. Moreover, grasshoppers, such as locusts, are important plant eaters in natural vegetation. Furthermore, application of pesticides to control grasshoppers and locusts has numerous negative consequences for the health of farm animals, human wellbeing, and wild animal populations (Branson et al., 2006).

Grass-feeding locusts species are common in rangeland and meadows, in which they act as livestock competitors. They commonly feed on Poaceae which belongs to legume family and is of great commercial value just like cereal grains and/or forage, but during the time of outbreaks, they may lead to significant damage to non-grass crops as described in Table 8.1 (Gall et al., 2019).

TABLE 8.1 Global Distribution of Locust Species and Their Grazing Interactions.

Species/subfamilies/Clade	Common name	Geographical distribution	Dietary preferences
	Tree and forb feeding locusts		
Calliptamus italicus **(Linnaeus, 1758)** **Calliptaminae Clade D**	Italian locust	Europe and Central Asia	NA
Anacridium melanorhodon **(Walker, 1870)** **Cyrtacanthacridinae Clade D**	Sahelian tree locust	North and Central Africa, Middle East	NA
	Grasses, forbs, and trees feeding locusts		
Schistocerca cancellata **(Serville, 1838)** **Cyrtacanthacridinae Clade D**	South American Locust	South America	Carbohydrate-biased P1: C2, (Overson et al., In prep)
Schistocerca gregaria flaviventris **(Burmeister, 1838)** **Cyratacanthacridinae Clade D**	Desert locust (southern race)	Southern Africa (N. B., typically non-swarming)	Carbohydrate-biased P1: C1, 25 (simpson et al., 2002)
Schistocerca gregaria gregaria **(Forskal, 1775)** **Cyrtacanthacridinae Clade D**	Dessert locust (northern race)	Southern Europe, North Africa, Middle East, Indian subcontinent	
Schistocerca interrita **(Scudder, 1899)** **Cyratacanthacridinae Clade D**	Peruvian locust	Peru	NA
Schistocerca piceifrons peruviana **(Lynch Arribalzaga, 1903)** **Cyratacanthacridinae Clade D**	Central American locust	Peru, Ecuador, Colombia, Venezuela, Panama, Trinidad and Tobago	NA
Dociostaurus maroccanus **(Thunberg, 1815)** **Gomphocerinae Clade B**	Moroccan locust	North Africa, Europe, Caucasus	NA
	Locusts feed on grasses		
Austracris guttulosa **(Walker, 1870)** **Cyrtacanthacridinae Clade D**	Spur-throated locust	Australia	NA
Nomadacris septemfasciata **(Audinet-Serville, 1883)** **Cytacanthacridinae Clade D**	Red locust	Sub-Saharan and Sothern Africa	NA

TABLE 8.1 *(Continued)*

Species/subfamilies/Clade	Common name	Geographical distribution	Dietary preferences
Patanga succincta **(Johansson, 1763)** **Cyrtacanthacridinae Clade D**	Bombay locust	India, South-East Asia	NA
Gomphocerus sibiricus **(Linnaeus, 1767)** **Gomophocerinae Clade D**	Siberian locust	Europe, Caucasus, North Asia	NA
Aiolopus simulatrix **(Walker, 1870)** **Oedipodinae**	Sudan plague locust	North and Central Africa	NA
Ceracris kiangsu **(Tsia, 1929)**	Yellow-spined bamboo locust	East Asia	NA
Chortoicetes terminifera **(Walker, 1870)** **Clade B**	Australian plague locust	Australia	NA
Locusta migratoria migratoria **(Linnaeus, 1758)** **Oedipodinae -Clade B**	Asian migratory locust	West and Central Asia, Eastern Europe	Carbohydrate-biased P1: C1, 4 (Chambers et al., 1995)
Locusta migratoria migratorioides **(Reiche and Fairmaire, 1849)** **Oedipodinae -Clade B**	African migratory locust	Mainland Africa and Atlantic Island, Madagascar, South and East Asia, Australasia, Tibetan plateau	
Locustana pardalina **(Walker, 1870)** **Oedipodinae Clade B**	Brown locust	Southern Africa	NA
Oedaleus asiaticus **(Bey-Bienko, 1941)** **Oedipodinae Clade B**	Mongolian locust	Northern Asia	Carbohydrate-biased P1: C2 (Cease et al., 2012)
Oedaleus senegalensis **(Krauss, 1877)** **Oedipodinae Clade B**	Senegalese locust	North Africa, Middle East	Carbohydrate-biased P1:C1, 6 (Le Gall, under review)

Source: Adapted and modified with permission from Cullen et al. (2017). © Elsevier.

8.3 EFFECT OF GLOBAL WARMING/CLIMATE CHANGE ON LOCUSTS POPULATION

Climate change is anticipated to raise earth's temperature and shift precipitation patterns (Cannon, 1998). Numerous research studies have been conducted on pest's behaviors to global climate change. Unfortunately, research on the consequences of global warming on locusts is limited and, to our understanding, only applies to the Oriental migratory locust species. Researchers compiled data of locust outbreaks and climate patterns in China for a thousands of years (957–1956) and discovered surprising findings: infestations were linked to chilled and humid periods, rather than warm and sunny ones, and water stress frequency (Zhang et al., 2009).

The geographic distribution of pests is simple to determine, especially in areas where population growth rate is favorable at a given point. It is a growing concern in recent times to talk about climate change and the rising threat of insects and pests attack to agricultural production and farming. Increased temperatures as a result of climate change could have a variety of effects on crop pest insect populations particularly locusts. Temperature influences insect growth, survival, and reproduction, as well as affecting the development and growth of their host plants. Temperature also influences the production of plant secondary compounds as well as the structural traits employed to protect against herbivores. As a result, temperature has the potential to have substantial repercussions for both insects and plants as shown in Figure 8.1 (Karthik et al., 2021).

Climate change has weakened plants' defense mechanisms against insects and pests, making them more vulnerable to harm. Initial invasion of *H. armigera* in cotton crop and pulses in Northern India is an example (Sharma, 2014). Rise in atmospheric CO_2 levels and global warming has also been discovered to reduce plant defenses against insect pests for instance in soybeans, the plant defense system signaling regulated by jasmonic acid (JA) that does not respond when CO_2 levels are high (Zavala et al., 2008). Locusts pests are evolving more overwintering phases and generations as a result of the rising growth trends caused by climate change.

Volatile Organic Compounds (VOCs) play a major role in a variety of pest–plant relationships, both positively (Pollination and seed dispersal) and negatively (e.g., herbivore defenses). Because plants generate higher levels of aromatic compounds in a dynamic environment, the air could become sufficiently pleasant as a consequence of global climate change, which can lead to have an impact on how flora interact with one another via competition and allopathic pathways, as well as how they defend against pests including insects, parasites, and viruses. Plants must produce higher amounts of VOCs for longer durations under warmer climates, affecting their natural functioning in insect–plant relations. Monoterpene emissions, for instance, are extremely sensitive to heat, increasing thrice with each 10°C rise in temperature (Constable et al., 1999).

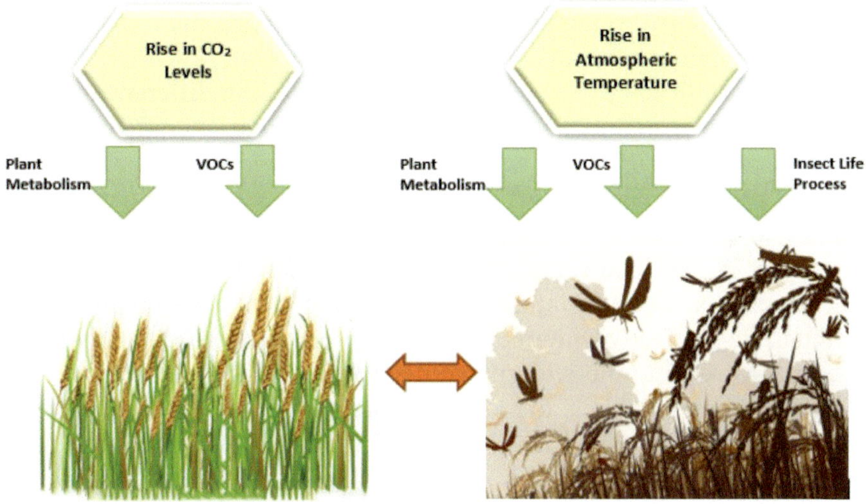

FIGURE 8.2 Effect of rise in temperature and CO_2 levels on crops, pests, and their interaction.

8.4 SUSTAINABLE AGRICULTURAL AND ECOLOGICAL PRACTICES—ADDRESSING LOCUST OUTBREAK

Several national and international groups/organizations are engaged in locust management and surveillance to limit agriculturists' susceptibility to locust plagues (Latchininsky, 2013). Scientific and geographical data on land-cover and habitat characteristics is needed to assess locust hazards and design preventative actions. Conventional land-monitoring approaches are clearly incapable of completing such a massive undertaking and providing credible spatiotemporal pests infestation record in terms of satellite image data, remote sensing seems to be crucial in tackling the magnitude of issue in locust dynamics. The earliest attempts at locust control using remote sensing were made in Africa and Australia in the 1970s and the 1980s (Hielkema, 1981).

Presently, satellite imaging covers the entire world and gives a plethora of information on what is going on the ground. Temperature, soil humidity, rainfall pattern, vegetation, and other data can be accessed. Furthermore, weather and climatic conditions are known to influence locust movement and appearance. In this regard, the fundamental concept is to provide responses at two phases: initially, to determine the impact of locusts and which areas require additional assistance, and then, to deliver replies at two phases, finally, to eliminate locusts by pinpointing the exact location at which a potential swarm could emerge (Akaninyene, 2020).

The area infested by the locust varies every year in different countries especially China, mainly due to global climate change, alterations in water systems, and anthropogenic activities, which tend to increase the conundrum of green prevention and control (GPC, like as environmental regulation and effective monitoring and management) of locust plague (Yao et al., 2017). For locust controlling, conventional quantitative strategies that depend on hybrid pinpoint investigations were incorporated. Such methods are ineffective, imprecise, and are prone to overlooking preventative actions. Due to the widespread distribution of the locust, these spot investigations are unlikely to provide efficient and targeted locust management. Remote sensing technology based on satellites can be helpful in evaluating extensive expanses of locust habitat. For more accurate and efficient locust control, a large-scale and high-precision method for locust area recognition is required (Mohammed et al., 2015; Ji et al., 2004).

Furthermore, as locust populations are generally found along rivers, lakes, and other bodies of water, using chemical pesticides in these regions can have major environmental consequences (Maute et al., 2017). Water pollution, soil pollution, biodiversity loss, and environmental harm are among major issues; pesticide use, however, may jeopardize the objectives of less-harmful pest management along with sustainable agriculture (Buckwell et al, 2014; Bais-Moleman et al., 2019). As per the research work performed by Puyun Yang, a pest control expert associated with National Agro-Tech Extension and Service Center, the following three measures have control efficiency of 70%, 20%, and 10%, respectively (personal communication). Environmental control, as opposed to chemical control, can provide protracted locust control by altering and lowering the reproductive conditions of locusts with least environmental damage. As a result, habitat modification and ecological control of locust areas play a significant role in locust population management and supporting sustainable agriculture (Zhao et al., 2020).

Policymakers and government agencies will be able to monitor agricultural trends across a vast area using satellite photography. Heeb et al (2019) proposed climate-smart pest management (CSPM) strategy to lessen the pest-induced damages to crops and agriculture and enhance food security (Fig. 8.3). This knowledge assists in disaster preparedness, such as severe weather events including drought and flooding, leading to increased profitability and productivity. When gathered over a period of time, this data can be used to forecast a whole state's crop yields, especially crop shortages. This vital data is required to appropriately plan for food scarcity and avoid famine by strategically placing supplies of food in places that are likely to face these difficulties. Instead of being in chaos or reacting after a crisis has occurred, these strategies can be used to forecast catastrophes (Alvarado-Serrano and Knowles, 2014).

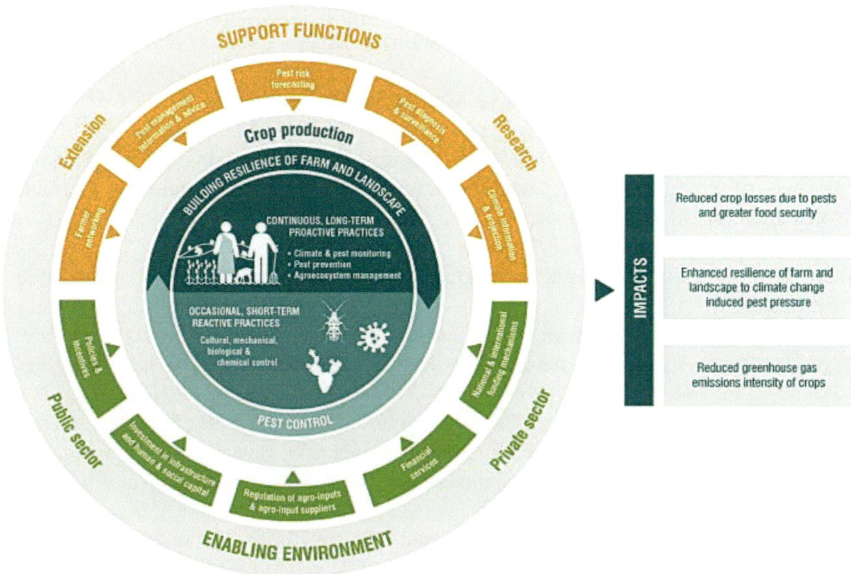

FIGURE 8.3 Climate-smart pest management (CSPM) is an integrative strategy aimed at making farmland more resilient to rising pest risks, mitigating climate change, and improving to food availability.

Source: Reprinted from Heeb et al., 2019. https://creativecommons.org/licenses/by/4.0/

KEYWORDS

- **locust**
- **plague**
- **global warming**
- **sustainable agriculture**
- **gregarious phase**
- **polymorphism**

REFERENCES

Alvarado-Serrano, D. F.; Knowles, L. L. Ecological Niche Models in Phylogeographic Studies: Applications, Advances and Precautions. *Mol. Ecol. Resour* **2014,** *14* (2), 233–248. https://doi.org/10.1111/1755–0998.12184

Anstey, M. L.; Rogers, S.; Ott, S.; Burrows, M.; Simpson, S. Serotonin Mediates Behavioral Gregarization Underlying Swarm Formation in Desert Locusts. *Science* **2009,** *323*, 627–630.

Areil, G.; Ayali, A. Locusts Collective Motion and Its Modelling. *PLOS Comput. Biol.* **2015,** *11* (12), e1004522. DOI: *10.1371/journal.pcbi.1004522*

Bais-Moleman, A. L.; Schulp, C. J. E.; Verburg, P. H.; Assessing the Environmental Impacts of Production- and Consumption-Side Measures in Sustainable Agriculture Intensification in the European Union. *Geoderma* **2019,** *338*, 555–567. https://doi.org/10. 1016/j.geoderma.2018.11.042.

Birch, L. C. The Role of Weather in Determining the Distribution and Abundance of Animal, 1957.

Branson, D. H.; Joern, A.; Sword, G. A. Sustainable management of insect herbivores in grassland ecosystems: new perspectives in grasshopper control. *Journal of Biological Sciences,* 56, 743–755. *doi: 10.1641/0006–3568(2006)56[743:SMOIHI]2.0.CO;2*

Buckwell, A.; Uhre, A. N.; Williams, A.; Poláková, J.; Blum, W. E. H.; Schiefer, J.; Lair, G. J.; Heissenhuber, A.; Schießel, P.; Krämer, C.; Haber, W. Sustainable Intensification of European Agriculture: A Review Sponsored by the RISE Foundation. Rural Investment Support for Europe, Brussels, 2014.

Cannon, R. J. C. The Implications of Predicted Climate Change for Insect PE\STS in the UK, with Emphasis on Non-Indigenous Species. *Global Change Biol.* **1998,** 4, 785–796.

Cease, A.; Elser, J. J.; Fenichel, E.; Hadrich, J. C. Living with Locusts: Connecting Soil Nitrogen, Locust Outbreaks, Livelihoods and Livestock Market. *BioScience* **2015,** *65* (6), 551–558.

Ceccato, P.; Cressman, K.; Giannini, A.; Trzaska, S. The Dessert Locust Upsurge in West Africa (2003–2005): Information on the Dessert Locust Early Warning System and the

Prospects for Seasonal Climate Forecasting. *Int. J. Pest Manage.* **2007,** *53* (1), 7–13. DOI: 10.1080/09670870600968826

Constable, J. V.; Litvak, M. E.; Greenberg, J. P.; Monson, R. K. Monoterpene Emission from Coniferous Trees in Response to Elevated CO2 Concentration and Climate Warming. *Global Change Biol.* **1999,** *5* (3), 252–267.

Crook, D. R.; Robinson, B. E.; Li, P. The Impact of Snowstorms, Droughts and Locust Outbreaks on Livestock Production in Inner Mongolia: Anticipation and Adaptation to Environmental Shocks. *Ecol. Econ.* **2020,** 177, 106761. https://doi.org/10.1016/j.ecolecon.2020.106761

Cullen, D. A.; Cease, A. J.; Latchininsky, A. V.; Ayali, A.; Berry, K.; Buhl, J. From Molecules to Management: Mechanisms and Consequences of Locust Phase Polyphenism. *Adv. Insect Physiol.*; Verlinden, H., Ed.; Elsevier: Oxford, 2017; pp 167–285. DOI: 10.1016/bs.aiip.2017.06.002

Gall, M. L.; Overson, R.; Cease, A. A Global Review on Locusts (Orthoptera: Acrididae) and Their Interactions with Livestock Grazing Practice. *Front. Ecol. Revol.* **2019,** *7*, 263. https://doi.org/10.3389/fevo.2019.00263

Heeb, L.; Jenner, E.; Cock, M. J. W. Climate-Smart Pest Management: Building Resilience of Farms and Landscapes to Changing Pest Threats. *J. Pest Sci.* **2019,** *92*, 951–969. https://doi.org/10.1007/s10340–019–01083-y

Hemming, C. F.; Popov, G. B.; Roffey, J.; Waloff, Z. Characteristics of Desert Locust Plague Upsurges, *Phil. Trans. R. Soc. London B, Biol. Sci.* **1979,** *287* (1022), 375–386. http://www.jstor.org/stable/241818

Hielkema, J. U. Desert Locust Habitat Monitoring with Satellite Remote Sensing: A New Technology for an Old Problem, *ITC J.* **1981,** *4*, 387–417.

Huffaker, C. B.; Dahlsten, D. L.; Janzen, D. H.; Kennedy, G. G. In *Ecological Entomology*; Huffaker, C. B.; Rabb, R. L., Eds.; Wiley Interscience: New York, 1984; pp 679–691; p 844.

Hunter, M. D. Effects of Elevated Atmospheric Carbon Dioxide on Insect Plant on Insect-Plant Interactions. *Agric. Forest Entomol.* **2001,** *3*, 153–159. https://doi.org/10.1046/j.1461–9555.2001.00108.x

Huntly, N. Herbivores and the Dynamics of Communities and Ecosystems. *Annu. Rev. Ecol. Systemat.* **1991,** *22*, 477–503.

Ji, R.; Xie, B. Y.; Li, D. M.; Li, Z.; Zhang, X.; Use of MODIS Data to Monitor the Oriental Migratory Locust Plague. *Agric. Ecosyst. Environ.* **2004,** *104*, 615–620. https://doi.org/10.1016/j.agee.2004.01.041

Karthik, S.; Reddy, M. S. S.; Yashaswini, G. Climate Change and Its Potential Impacts on Insect-Plant Interactions. *Global Warm. Clim. Change.* DOI: 10.5772/intechopen.98203

Latchininsky, A. V. Locusts and Remote Sensing a Review. *J. Appl. Remote Sens.* **2013,** *7* (1), 075099–1. DOI: 10.1117/1.JRS.7.075099

Lecoq, M. Dessert Locust Management: From Ecology to Anthropology. *J. Orthoptera Res.* **2005,** *14* (2), 179–186. https://www.jstor.org/stable/3657126

Lima, M. Locust Plagues, Climate Variations and the Rhythms of Nature. *Proc. Natl. Acad. Sci. USA* **2007,** *104* (41), 15972.15973.

Lockwood, J. A. Locust: The Devastating Rise and Mysterious Disappearance of the Insect That Shaped the American Frontier; Perseus, New York, 2004.

Maute, K.; French, K.; Story, P.; Bull, C.M.; Hose, G. Short and Long-Term Impacts of Ultra-Low-Volume Pesticide and Biopesticide Applications for Locust Control on Nontarget Arid Zone Arthropods. *Agric. Ecosyst. Environ.* **2017,** *240*, 233–243. https://doi.org/10.1016/j.agee.2017.02.024

Mohammed, L.; Diongue, A.; Yang, J. T.; Bahia, D. M.; Michel, L. Location and Characterization of Breeding Sites of Solitary Desert Locust Using Satellite Images Landsat 7 ETM+ and Terra MODIS. *Adv. Entomol.* **2015,** *3*, 6–15. https://doi.org/10.4236/ae.2015.31002.

Nevo, D. The desert locust *Schistocerca gregaria*, and its control in the land of Israel and the near east in antiquity, with some reflections on its appearance in Israel in modern times, *Phytoparasitica*, **1996,** 24, 7–32.

ObotAkaninyene. Locust Invasion Prediction for Future Food Security and Sustainable Agriculture in East Africa Case Study SmartAfriHub Hackathon Challenge 4. *J. Res. Agric. Anim. Sci.* **2020,** *7* (5), 20–29.

Pener, M. P.; Simpson, S. J. Locust Phase Polyphenism: An Update. *Adv. Insect Physiol.* **2009,** *36*.

Pener, M. P.; Yerushalmi, Y. The Physiology of Locust Phase Polymorphism: An Update. *J. Insect Physiol.* **1998,** 44 (5–6), 365–377. doi:10.1016/s0022–1910(97)00169–8

Peng, W.; Ma, N. L.; Zhang, D.; Zhou, Q.; Yue, X.; Khoo, S. C.; Yang, H.; Guan, R.; Chen, H.; Zhang, X.; Wang, Y.; Wei, Z.; Suo, C.; Peng, Y.; Yang, Y.; Lam, S. S.; Sonne, C. A Review of Historical and Recent Locust Outbreaks: Link to Global Warming, Food Security and Mitigation Strategies. *Environ. Res.* **2020,** *191*, 110046. https://doi.org/10.1016/j.envres.2020.110046

Popov, G. A. Dynamics of Locust Population and Their Harmfulness. In *Locusts: Ecology and Control Methods*; E. M. Shumakov, Ed.; All-Union Institute for Plant Protection, Leningrad: USSR, 1987; pp: 12–29.

Sharma, H. C. Climate Change Effects in Insects: Implications for Crop Protection and Food Security. *J. Crop Improv.* **2014,** *28*, 229–259.

Showler, A. T. Locust (Orthoptera: Acrididae) Outbreak in Africa and Asia, 1992–1994: An Overview. *Am. Entomol. 1995*, *41* (3), 179–185. https://doi.org/10.1093/ae/41.3.179

Simpson, S. J.; Raubenheimer, D.; Behmer, S. T.; Whitworth, A.; Wright, G. A. A Comparison of Nutritional Regulation in Solitarious and Gregarious-Phase Nymphs of the Desert Locust *Schistocerca gregaria. J. Exp. Biol.* **2002,** *205*, 121–129.

Song, H. Density-Dependent Phase Polyphenism in Nonmodel Locusts: A Mini Review; *Hindawi Publishing Coopooration* **2010, 2011,** 741769. DOI:10.1155/2011/741769

Stark, R. W.; Burnell, D. G.; Neuenschwander, L. F.; Stock, M. W.; Nathanson, R. A. Dynamics of Parasite Guilds and Insect Herbivores in Forest Successional Stages, 1985; pp 10–23.

Steedman, A. *Locust handbook*; Natural Resources Institute: Chatham, 1990.

Toleubayev, K.; Jansen, K.; Huis, A. V. Locust Control in Transition: The Loss and Reinvention of Collective Action in Post-Soviet Kazakhstan. *Ecol. Soc.* **2007,** *12* (2), 38. http://www.ecologyandsociety.org/vol12/iss2/art38/

Topaz, C. M.; D'Orsogna, M.; Edlestein-Keshet, L.; Bernoff, A. J. Locuts Dynamics: Behavioral Phase Change and Swarming. *PLOS Comput. Biol.* **2012,** *8* (8), e100https://doi.org/10.1371/journal.pcbi.1002642

Uvarov, B. *Grasshoppers and Locusts*, Vol. 1; Cambridge University Press: Cambridge, 1966.

Van Huis, A. Locust Control by Early Identification of Breeding Sites. *Encyclopedia Pest Manage* **2007,** *1* (1), 1–3.

Van Huis, A.; Cressman, K.; Magor, J. I. Preventing Desert Locust Plagues: Optimizing Management Interventions. *Entomol. Experimentalis et Applicata* **2007,** *122*, 191–214.

Yao, X.C.; Zhu, D.H.; Yun, W.Y.; Peng, F.; Li, L.; A Web GIS-Based Decision Support System for Locust Prevention and Control in China. *Comput. Electron. Agric.* **2017,** *140*, 148–158. https://doi.org/10.1016/j.compag.2017.06.00

Zavala, J. A.; Casteel, C. L.; Delucia, E. H.; Berenbaum, M. R. Anthropogenic Increase in Carbon Dioxide Comprises Plant Defense Against Invasive Insects. *Proc. Natl. Acad. Sci. USA* **2008,** *105*, 5129–5133.

Zhao, L.; Huang, W.; Chen, J.; Dong, Y.; Ren, B.; Geng, Y. Land Use/Cover Changes in the Oriental Migratory Locust Area of China: Implications for Ecological Control and Monitoring of Locust Area. *Agric. Ecosyst. Environ.* **2020,** *303*, 107110.

CHAPTER 9

Locust Outbreaks in India and in the Cold Arid Region of Ladakh and Their Management

RAYEES AHMAD[1] and BARKAT HUSSAIN[2]

[1]*Entomology Research Lab, Department of Zoology, University of Kashmir, Srinagar, Jammu and Kashmir, India*

[2]*Division of Entomology, Department of Horticulture, SKUAST-K, Srinagar, Jammu and Kashmir, India*

ABSTRACT

Locusts has been considered one of the dreadful and destructive pests since the dawn of civilization, even mentioned in Holy Books and always considered world's devastating pest attacking all crops, even all types of grasses, pastures, and trees. The devastation caused by the locusts is always alarming and threatening, leading to plagues and the losses caused by the locusts in India alone are more than thousand billion dollars (US $). To prevent losses caused by the locusts, Government of India established locust warning organization to mitigate the control measures in advance and it is the oldest organization in the globe to prevent locust invasion. The historical locust outbreaks of different intensities in India (1812, 1821, 1843–1844, 1863–1867, 1869–1873, 1876–1881, 1889–1898, 1900–1907, 1912–1920, 1926–1931, 1940–1946) had been reported and the losses have not been quantified collectively on all crops, pastures, trees,

Locust Outbreaks: Management and the World Economy. Umair Riaz, Khalid Rehman Hakeem, (Eds.)

and the amount of money incurred by the government on procurement of huge pesticides for managing these locust outbreaks and other health hazards inflicting on human population and livestock. Recent outbreaks have been reported in the fragile environment of Ladakh. However, the big locust outbreak was reported in 2010 in India. Thirteen locust plagues were reported since 1964–1997, five between1997 and 2010, but these were controlled according to Locust Control Organization. But, no major outbreaks were reported during 2010–2018. The recent locust outbreaks of 2019 and 2020 have significantly affected badly major crops of India. The recent locust outbreak of India even reached to Delhi and even UP and affected human activities and nuisance to welfare of urban population. Ladakh is the second highest and coldest region of the World and the outbreak of locusts in Zanskar during 2006 is a cause of concern to fragile environment and 90% damage had been reported and since then small outbreaks in other parts of Ladakh have been reported. The management of locusts in India is elaborated briefly in this chapter with recent advances by using drone technology to eradicate the locust swarms aerially with precision, accuracy, and in a shorter period of time with ease. The damage caused by the locusts is so severe and consuming the livelihood of one-tenth of the world's population.

9.1 INTRODUCTION

Locusts (Orthoptera: Acrididae) are short-horned grasshoppers, having more than one phase in their life—solitary and gregarious. Locusts and grasshoppers are an important source of food for many animals including spiders and birds (Wiens, 1973; Joern, 1986; Oedekoven and Joern, 2000; Schmitz, 2008; Sokol-Hessner and Schmitz, 2002; Belovsky and Slade, 2000), and also humans (Stoops et al., 2016; Osimani et al., 2017). After crickets, locusts are the second most commonly eaten insects (Osimani et al., 2017). Around 6800, acridid grasshopper species are known so far (Cigliano et al., 2017). Among all, there are 19 species in the world currently considered locusts (Cullen et al., 2017), and only 04 species are found in India. Under many circumstances, the solitary phase of locusts changes into a gregarious phase by changing their behavior, body-color, and habit. Due to the high densities of locusts, they compete with live stocks for forage and become highly problematic. Locusts when exposed to specific environmental conditions change their behavior and develop into a gregarious phase and start swarming (Cullen et al., 2017; Pener, 1991). Their massive swarm is sweeping across many countries and the new outbreaks are worse than the previous ones. During swarming, billions of locusts accumulate per square kilometer, move toward greenery, and eat everything that came in their path. Because of their huge population density per square kilometer, it becomes very difficult to control them. Only developed countries can manage some controlling tactics as it needs advanced aerial sprayers as well as coverage of the huge land area. Management of locusts with mechanical, chemical, baiting, dusting is not up to the mark as these methods have many negative side effects. Another management strategy in the form of making delicious food from locusts because of their delicacy and edible nature is also promising. Locusts are very rich in proteins and other essential nutrients, hence these are used as a food in many Asian, African, and Middle Eastern countries. It is better to get natural proteins by eating crispy locusts than to allow them to harm our agriculture sector. The food demand to the hungry people and to reduce crisis and serious threat to crops, their economy and livelihood will be fulfilled to the great extent.

9.2 BEHAVIOR AND LIFE CYCLE OF LOCUSTS

From historical times locusts have been revered and feared in almost every country. Locusts look like grasshoppers as both have big hind legs for hopping or jumping. Locusts behavior is sometimes completely different from the grasshoppers. Locusts undergo several phases before attaining maturity as adults. During dry conditions, solitary locusts aggregate together in the patchy areas of land with remaining vegetation. Due to sudden aggregation and overcrowding, serotonin is released in their central nervous system that makes locusts more sociable, with a more varied appetite, and promotes rapid moments. As soon as the rainy season and vegetation return, the perfect environmental conditions are now available to create a perfect storm. During these circumstances, the solitary lifestyle is completely shifted to the gregarious phase. Color, body shape, other physical traits, and behavior are changed completely during the gregarious phase. Their brains and endurance increase to a large extent. These transformations in their behavior and physical traits can eventually be reversed, or these can persist and be passed to the next generation. On hatching, locusts emerge wingless as apterygote nymphs, which can be either solitary or gregarious.

9.3 SWARMING OF LOCUSTS IN MANY COUNTRIES

Locusts are short-horned grasshoppers and are termed the oldest migratory pests in the world. They form swarms of adults where they breed and feed indiscriminately on any green vegetation, coming in their way. They swarm typically in motion and can cover very large distances. The swarming of locusts is highly mobile and can be very dense. One such swarm can have around 80 lakhs of locusts. Some species can travel 81 miles or more a day and can stay in the air for a long period with non-stop trips across many countries. Swarming is to some extent a continuous process as new swarms or bands are formed again and again in new fields. In 1954, a swarm of locusts start flying from Northwest Africa to Great Britain, while in 1988, another swarm flew from West Africa to the Caribbean that covered 3100 miles in just 10 days. Massive swarms of locusts are sweeping across the Middle East, the Horn of Africa, and South Asia. In the Horn of Africa, around 1 million lands are under threat in almost eight different countries. Some current outbreaks are the worst in recent decades and are responsible for the economic loss, livelihood and pose a serious threat to the standing cash and other crops. Foreign experts especially from the United States have warned it to be posing a severe risk to Indian agriculture. Plagues of locusts devastate crops and cause major agricultural damage in many regions since prehistoric times and can lead to famine and starvation and they still wreak havoc today. African regions are one of the most destructive in subsistence farming due to locust swarming. The locust swarm in Senegal, West Africa, is considered second only to drought in damaging agricultural productivity (D'Alessandro et al., 2015). It has been clearly mentioned in both the Al-Quran and the Bible that locusts are one of the devastating plagues of many regions of Egypt (El-Mallakh and El-Mallakh, 1994).

At the global level, the following regions are or can be highly affected by the attack of locust swarms as per FAO (Table 9.1). The havoc started with the onset of heavy rainfall in East Africa and the Arabian Peninsula. The most predominant species and those reported from India creating havoc in many parts of the world are enlisted in Table 9.2.

TABLE 9.1 Countries and Cities with Serious Threat of Locust Swarm.

Locust attack regions	Countries or cities with locust swarming
East Africa	South Sudan, Uganda, Kenya, Somalia, Ethiopia,
West Africa	Chad
Arabian Peninsula	Yemen, Saudi Arabia, Oman, UAE,
South West Asia	Iran, Pakistan, India

TABLE 9.2 The Most Prominent Locust Species and Their Presence in India, Responsible for Havoc.

S. No.	Species	Scientific name	Family	Reported from Indian
01	Desert locust	*Schistocerca gregaria*	Cyrtacanthacridinae	+
02	Bombay locust	*Nomadacris succincta*	Oedipodinae	+
03	Migratory locust	*Locusta migratoria*	Oedipodinae	+
04	Tree locust	*Anacridium melanorhodon*	Cyrtacanthacridinae	+
05	Italian locust	*Calliptamus italicus*	Calliptaminae	-
06	Moroccan locust	*Dociostaurus maroccanus*	Gomphocerinae	-
07	Red locust	*Nomadacris septemfasciata*	Cyrtacanthacridinae	-
08	Brown locust	*Locustana pardalina*	Oedipodinae	-
09	South American locust	*Schistocerca cancellata*	Cyrtacanthacridinae	-
10	Australian locust	*Chortoicetes terminifera*	Oedipodinae	-
11	Peruvian locust	*Schistocerca interrita*	Cyrtacanthacridinae	-
12	Central American locust	*Schistocerca piceifrons*	Cyrtacanthacridinae	-
13	Spur-throated locust	*Austracris guttulosa*	Cyrtacanthacridinae	-
14	Siberian locust	*Gomphocerus sibiricus*	Gomphocerinae	-
15	Sudan plague locust	*Aiolopus simulatrix*	Oedipodinae	-
16	Yellow-spined bamboo locust	*Ceracris kiangsu*	Oedipodinae	-
17	Yellow-winged locust	*Gastrimargus musicus*	Oedipodinae	-
18	Mongolian locust	*Oedaleus decorus*	Oedipodinae	-
19	Senegalese locust	*Oedaleus senegalensis*	Oedipodinae	-

9.4 LOCUST SWARMS IN INDIA

In India, four locust species are creating plague from time to time. These locusts can travel in large swarms with a speed up to 150 km per day, depending on the speed of the wind as well. They are causing direct as well as indirect losses. Direct losses include their feeding pattern and the indirect one included the breakage of the plant's twigs, parts, as well as the whole plant with the excessive weight of accumulated gregarious locusts. Hence, the attacks lead to starvation and famine by causing massive agricultural damage. Every part of the plant is equally affected indiscriminately, such as seeds, leaves, fruits, barks, flowers, and tips or new emerging buds. It has been said that a small swarm of locusts has an appetite which equals the appetite of 3500 people, 20 camels, and six elephants.

Locust swarm was the fourth great blow to the Indian economy after coronavirus, cyclones, and floods, especially during the year 2020. The agricultural sector is the primary sector in India but is greatly shaken to the ground by the locust storm. The year 2020 has brought a lot of miserable suffering to the inhabitants of India. On one side coronavirus was engulfing the lives indiscriminately and on the other side, India has to deal with the huge loss due to a locust swarm attack, which returned almost after three decades. Locust swarm during 1926–1931, 1940–1946, 1949–1955, and 1959–1962 caused damage worth rupees 10 Crores, 02 Crores, 02 Crores, and 50 Lakhs, respectively. Although farmers are using many manual methods like the beating of drums, smoke emission, and chemical spray to avoid the swarm, still huge damage was seen especially from Rajasthan, Gujrat, Madhya Pradesh, Uttar Pradesh, Maharashtra, Punjab, and Delhi. The main reason behind locust swarming is the occurrence of cyclone Amphan, which came in May 2020.

India faces an average of 10 attacks of locusts per year. During 2020, the most attractive seasons for locust swarming, due to excess rainfall and vegetation in many states. Excess rainfall is also an important factor for the excessive oviposition of locusts. The onset of monsoon, as well as change in wind direction, triggers locusts to move toward India from Saudi Arabia. They can cross the red sea and fly toward Iran and Pakistan. In the past 26 years, the locust swarm of 2020 is termed the worst one. Five states have been recently affected by the locust swarm and made the headlines with its early arrival in India. The states included Maharashtra, Punjab, Haryana, Rajasthan, and Madhya Pradesh. The reason behind early locust swarming in India was longer monsoon than usual across the Indo–Pakistan border and frequent cyclones in the Indian Ocean. According to FAO, around 80 million adult locusts aggregate within a distance of one square kilometer and can eat as much food as 35,000 people in terms of weight in a single day. Nagaur, Jaipur, Dausa, Bangra Magarpur, Wardha, Jhansi, Amravati,

Malhargarh are the various cities and districts that are extremely affected by the outbreak of locust swarms. The following species mention the state of India amid current locust swarm's:

1. *Schistocerca gregaria flaviventris* (Burmeister, 1838) and *Schistocerca gregaria gregaria* (Forskal, 1775) (Desert locust):

These notorious species are found in Southern Africa and follow the wind patterns to cover a vast stretch from Saudi Arabia to the deserts and other states of India such as Maharashtra, Gujrat, Punjab, Madhya Pradesh, and Uttar Pradesh. Desert locusts are commonly found in North Africa, the Middle East, and South Asia. Several countries are covered during the swarming of locusts and inhabit an area of about six million square miles. 30–60 countries can be covered and affected by this notorious species during swarming and cover 1200 square kilometers (Mohamed, 2007). The covered area can be represented as the fifth of Earth's land surface. The swarm of these desert locusts can also be around 460 square miles in size and pack between forty and eighty million locusts into less than half a square mile. Desert locusts are also notorious for the indiscriminate feeding of any green plants that come in their way and threaten the economic livelihood of humans and cause an indirect effect on the grazing animals as well. Many locust species originate in such areas used for livestock grazing (Branson et al., 2006). Desert locusts are voracious feeders and can eat about a million pounds of plants every day. It has been said that a swarm of the size of Paris can eat the same amount of food in one day as half the population of France. The plaques caused by this locust were started as early as 2420 BC in the tombs of Egypt (Nevo, 1996), and even remain a threat to modern times. The other plagues were seen in the years 1962, 1963, 1965, 2003, 2004, 2005, and 2006. According to Belayneh (2005), the last plague of 2003–2005 was one of the dreadfuls that caused 80–100% crop loss, especially in Sub-Saharan Africa (Brader et al., 2006). Schistocerca gregaria is often considered the most dangerous migratory locust at the global level due to its vast reach and significance (Steedman 1990).

2. *Nomadacris succincta; Patanga succinate* (Johansson, 1763) (Bombay locust):

The Bombay locust is distributed throughout the Asia and Middle East regions and inhabits grassy plains of Asia up to about 1500 m. A large variety of crops have been damaged by the Bombay locusts such as banana, bamboo, beans, coconut, cashew, castor, citrus, cowpea, cucumber, fig, ginger, guava, mango, mulberry, mustard, sugarcane, sweet potato, palm, tobacco, and tea (Roffey, 1979). In Rajasthan, this locust is tremendously feeding on a sedge, Cyperus tuberosus. Swarming in India is started in the forests of the Western Ghats during the winter disperses through north or east after rising in temperature. Swarming by this locust was recorded from 1787 to 1796 and from 1901 to 1908. In Thailand, these locusts are considered one of the worst agricultural pests that damage 8000 acres of maize crop worth 50,000 Euros, during 1961–1963 crop seasons. It is also consumed by the inhabitants of Thailand where it is considered the most popular insect food item that is commercially produced for human consumption (Phiriyangkul et al., 2015). In Malaysia, this notorious pest mainly attacks rice fields, as they initially feed on the grasses.

3. *Locusta migratoria migratoria* (Linnaeus, 1758); *Locusta migratoria migratoriodes* (Reiche and Fairmaire, 1849) (Migratory locust):

The migratory locust is abundant across the entire Tropical Eastern and Temperate hemispheres. These are the most widely distributed grass-hopper species in the world. Locust storm by the Asian migratory locust was reported from China from 200 BC and has a typical association with drought and flood events (Stige et al., 2007; Tian et al., 2011). In Australia, this particular species causes excessive damage and an outbreak occurred in the Central Highlands of Queensland in 1973–1976 (Farrow, 1979). Other countries highly affected by this pest are USSR, Uzbekistan, and Kazakhstan (Latchininsky, 2013). In 1946, over 1 million ha of vegetation land has been affected by these locusts that are strong fliers and cover over 1000 km at a time. The main plants to be damaged during the migratory locust storm were banana, pineapple, palms, grasses such as Panicum sp, Artemisia sp., Polygonum sp., and Phragmites communis, and other cereal summer crops. Nymphs of L. m. migratoria are polyphagous, their 1st and 2nd instar are mostly feeding on grasses and the 3rd ones are later shifted to barley and wheat crops (Qing and Chunxian, 2008).

4. *Anacridium melanorhodon* (The Sahelian tree locust): Cyrtacanthacridinae

The A. melanorhodon is commonly found in drought and arid habitats. It is referred to as Sahelian tree locust because of its distribution in the Sahelian zone from Cape Verde to Ethiopia as well as some parts of the Middle East. It is nocturnal and usually absorbs water from frass during hot and dry conditions (Abushama, 1970). It is a voracious feeder and its feeding behavior is continuously stimulated by the phenolic compounds that it prefers to eat. The A. melanorhodon is found in India, Sudan South Sudan, Oman, North America, and Ethiopia. Grasses, herbs, shrubs, and trees are equally infested by the different stages of this locust. Some host trees are gum Arabic, citrus, apple, palm, mango, olive, millet, sorghum, rice, maize, cassava, cotton, and orange trees. Gum production is extremely affected by this pest in Sudan and South Sudan, where Acacia senegal is one of the main crops produced in the traditional rain-fed agricultural sector (Haroon et al., 2011). The A. melanorhodon destroys almost 50 percent of gum Arabic crop that reduced from 273.9 to 93.8 kg/ha.

9.5 MANAGEMENT OF LOCUST SWARM IN INDIA

Locust warning organization has been established in India with headquarter at Jodhpur. Its main function is to provide information and help to the state Govt. in controlling invasions. The following states have managed the swarm in the following ways.

9.5.1 RAJASTHAN

Rajasthan state shares its borders with its neighboring country Pakistan, from where locust swarm invaded. The locust outbreak badly hits the 12 districts of Rajasthan, which compels the Government to plan some effective controlling measures. Spraying of pesticides with the help of drones in remote access areas, implementation of the Prime Minister's crop insurance scheme, and preparation of contingency plans has been developed so far by the Rajasthan Government to tackle the locust outbreak.

9.5.2 UTTAR PRADESH

Locusts spoil the standing crops and trees to a great extent in many districts of Uttar Pradesh. Locust storm arrived in Uttar Pradesh through the Jhansi district. Swarming of locusts was immediately controlled by the UP Government, and became successful in eliminating 70% of the pest population with the help of the Fire Service and Agriculture department. Spraying of pesticides, burning the straw, and machinery was set ready to tackle any further attack.

9.5.3 MAHARASHTRA

Locusts mainly attack Chikhaldara, Morshi, Katol, Warud, and Ramtek regions of Maharashtra. Locusts mainly attack green vegetation and cause direct damage by eating and indirectly by overweight on branches and twigs of crop plants. The State Agriculture Department, along with other official members and locals, has driven away and fought against locust swarm by beating drums, and spraying disinfectants. During the crop season 2020, a total of 25 drones and more than 70 vehicles mounted with insecticide-spraying machines were deployed to attack the locust swarm in Maharashtra.

9.5.4 GUJARAT

Although not much heavy damage was seen here in the Gujarat state of India, however, the Gujarat administration had already warned the state regarding the upcoming worst situation caused by the locust storm. Knowledge sharing to local farmers about the handling of situations, pesticide stocking, spraying machines mounted on tractors and other vehicles, and establishment of water pumps were adopted by the Government of Gujarat.

9.5.5 DELHI

Playgrounds, sugarcane fields, and metro stations were badly hit by the plague of 2020, in Delhi. A swarm of about two kilometers had spread over Delhi and moved toward Gurgaon and covered the sky. Delhi administration also took immediate steps to overcome the locust attack. Spraying of pesticides with advanced machinery and irrigation facility to agricultural lands was made possible to tackle any future attack.

9.6 LOCUST SWARM IN LADAKH

Trans Himalayan Ladakh, a Union Territory of India, is known for the incredible topography that comprises hilly terrains, the lush green grasslands, and the high-altitude peaks. In Ladakh, various fruits and vegetables like apricot, long melon, watermelon, peach, lettuce, cucumber, capsicum, pumpkin, etc., are cultivated in areas like Darchiks, Batalik, Hardas, Shilikchay, Saspole, Badgam, and Chanigund. The people of Ladakh are grappling with the menace of locusts. It has been found that the Locusta migratoria causes an extensive loss of crops in Ladakh. L. migratoria have migrated from China along the banks of the Indus River to Ladakh. Local nomads claim that 2 years back near the China border at Demchong, locusts have entered inside Ladakh. Due to excessive rainfall in China in 2020, an ideal environment has been provided across the border, resulting in the worsened crop loss in Ladakh. The crops of Changthang and its adjoining areas are very much susceptible to infestation by locust swarming, where 80% of crops and 70% of pasturelands have already been ruined by swarms of locusts. Preventive and controlling strategies were immediately suggested by a team of entomologists keeping in view the onslaught of locusts. Spraying of pesticides in the form of chlorpyriphos was suggested by the experts, but local villagers of Demchong, Koyul, Skakjung, and Rongo have denied using the spray to kill the pests due to their religious sentiments. On the other side, environmentalists are also worried about the excessive use of pesticides, and it kills the nontarget species as well. However, pesticides were sprayed to reduce the locust population. Changsheng's fields are later seen to be filled with the insecticide-laden carcasses of locusts as well as other animals feeding on them. From Ladakh, the swarm of locusts spread further to warmer regions of the country such as Jammu and Himachal Pradesh. Following the onslaught, the Ladakh administration should have to prepare fully in order to tackle the situation in the future as soon as the swarm of locusts begins. Instead of spraying pesticides, the concept of IPM should be adopted for the maximum benefit as well as in reducing pollution and other side effects.

9.7 PREVENTIVE MEASURES

It is very difficult to control a plague of locusts due to the large number and large occupied area by locusts. Another reason for its difficult prevention is the lack of economical sources in an affected country. However, some preventive measures should be adopted and are very fruitful. Proper information of weather patterns and by checking historic records to identify the regions greatly affected by the locust swarm and spray those identified areas with chemicals to reduce their emergence and hence control their population. Excess temperature is also one of the leading causes of locust swarm and scientists are worried about the plagues that will emerge and create worse conditions in the warming regions of the world. High temperatures also cause cyclones to emerge in eastern Africa and the Arabian Peninsula, where the swarming of locusts thrive.

9.8 PRECAUTIONARY MEASURES

Controlling strategies are not easy. Farmers should be fully aware of the life cycle of specific locust species to control the upcoming swarm. Locusts rest during night and evening hours on the ground, and farmers are advised to form groups and monitor their fields during evening hours. Dig large ditches around their fields and beating drums or play any loud instruments to flee these locusts under rest. Use of botanicals such as neem-based formulations, Azadirachtin 1500 ppm 5 mL per L should be sprayed on crops, especially during late evening and night hours.

9.9 MANAGEMENT STRATEGIES FOR LOCUST CONTROL

Effective management strategies of locust upsurges require the proper management strategies both for robust monitoring of adult and lower nymph stages. Local administration incorporation with international effort could help to tackle the locust swarm effectively (Lockwood et al.,

2001). Various management strategies are enlisted under the following subheadings:

9.9.1 PHYSICAL AND CULTURAL CONTROL

Physical and cultural control strategies include:

1. Use of fire strategies like lighting fires or making noise to prevent swarms from settling in crops. (Farmers can install tape recorders with mounted pen drives having stored some harsh sounds)
2. Digging trenches and driving hoppers into trenches and burning, drowning, and crushing them.
3. Beating or trampling on the hoppers.
4. Digging up egg pods or ploughing fields infested with egg pods.
5. Use of flame throwers.

9.9.2 CHEMICAL CONTROL

The chemical control strategy includes:

1. Spraying Chloropyriphos 20EC @ 1 mL/L of water or Malathion 50 EC@ 1 mL/L of water in the affected areas.
2. The most efficient and commonly used method for locust control is the use of a technique known as Ultra Low Volume (ULV) spraying which is the technique of using a much smaller volume of spray liquid by air, that is, only 0.5–1.0 L per hectare is recommended for locust control. But keeping in view the Ladakh climatic condition the formulation should be oil-based.
3. ULV sprayers can be carried by an operator (Portable sprayer), mounted on a four-wheeled drive vehicle (vehicle-mounted sprayer), or an airplane or helicopter aircraft-mounted sprayer.
4. Wherever locust swarm is small and has a very low density, then isolated controlling measures can be useful to control the small population.

Although pesticide spay is very toxic and has various side effects, their usage is required during the swarming of locusts. Some insecticide sprays along with their chemical dosage per hectare of vegetation are listed in Table 9.3.

TABLE 9.3 Various Pesticide Sprays Along with Their Dosage Per Hectare of Vegetation.

S. No.	Technical	Toxicity	Dosage/acre	Mode of action	Chemical name
01	Bendiocarb 80%WP	Extremely toxic	50g	Systemic	2,2-Dimethyl-2*H*-1,3-benzodioxol-4-yl methylcarbamate
02	Chloropyriphos 50%EC	Highly toxic	194 mL	Systemic and Contact	O,O-diethyl O-(3,5,6-trichloro-2-pyridinyl) phosphorothioate
03	Chloropyriphos 20%EC	Highly toxic	0.50 L	Systemic and Contact	O,O-diethyl O-(3,5,6-trichloro-2-pyridinyl) phosphorothioate
04	Deltamethrin 2.8%EC	Highly toxic	182 mL	Systemic and ingestion	(*S*)-Cyano(3-phenoxyphenyl) methyl (1*R*,3*R*)-3-(2,2-dibromoethen-1-yl)-2,2-dimethylcyclopropane-1-carboxylate
05	Deltamethrin 1.25%ULV	Highly toxic	80 mL	-	-
06	Lamda cyhalothrin 5%EC	Highly toxic	161 mL	Contact and systemic	(R)-α-cyano-3-phenoxybenzyl (1S, 3S)-3-[(Z)-2-chloro-3, 3, 3-trifluoropropenyl]-2, 2-dimethylcyclopropanecarboxylate and (S)-α-cyano-3-phenoxybenzyl (1R, 3R)-3- [(Z)-2-chloro-3, 3, 3-trifluoropropenyl]-2, 2-dimethylcyclopropanecarboxylate
07	Lamda cyhalothrin 10%WP	Highly toxic	80g	–	–
08	Fipronil 5%SC	Highly toxic	50 mL	Contact and Ingestion	5-amino-[2,6-dichloro-4-(trifluoromethyl) phenyl]- 4-[(1R,S)-(trifluoromethyl)sulfinyl]-1H-pyrazole-3- carbonitrile
09	Fipronil 2.92%EC	Highly toxic	87 mL	–	–
10	Diflubenzuron 25%WP	Highly toxic	97g	Contact and systemic	*N*-[(4-chlorophenyl) carbamoyl]-2,6-difluorobenzamide
11	Malathion 50%EC	Moderately toxic	0.70l	Contact	diethyl (dimethoxy phosphinothioyl) thiobutanedioate
12	Malathion 25%WP	Moderately toxic	1.50kg	–	–

Source: ICRISAT (2020).

9.9.3 BAITING

Baiting simply means food or any other substitute used as a lure to trap or kill the target pest against whom work is done. To control a swarm of locusts, maize meal, wheat bran, rice bran, and cottonseed husk are some of the best carriers for locust bait. The carrier is mixed with the required concentration of insecticide and the appropriate value is 20:1 (Carrier: Insecticide). For instance, hoopers as well as other stages can be controlled by the bait that can be prepared with the mixing of 20 kg of wheat or other bran and 1 kg of 1% bendiocarb dust or 3 mL of fipronil 5%SC. The bait is only successful when much bare ground and little vegetation is present in the way of marching bands. Another limitation of baiting is that the hoopers are not attracted toward it from a distance; instead, we have to put it in their way. Baits should be sprayed thinly to cover the maximum area and it is possible only by throwing it high in the air and letting it drift with the wind.

9.9.4 DUSTING

A speck of fine dust consisting of sawdust, sand, or soil impregnated with bendiocarb as a killing agent is applied by hand for better results. The appropriate ratio is the mixing of one handful of an insecticide dust to four handfuls of dry sand or silt. First instar hoppers, particularly marching through dense low vegetation on which they feed, can be controlled by applying the insecticide dust such as Fenvalrate 0.4Dp, Quinalphos 1.5Dp, and Malathion 5Dp.

9.10 CONCLUSIONS

India has to fight another battle against the swarms of locusts besides the COVID-19 pandemic. The swarm of locusts entered into India through western states and have already made their way into many states of India such as Rajasthan, Gujarat, Madhya Pradesh, Maharashtra, and Uttar Pradesh. Keeping in view the favorable climatic conditions, locusts are now multiplying at an astonishing 400 times than their usual rate. The western part of India is under the great threat of locust attack that can lead to a major problem of food security of the country if not managed properly. Not only at the state level but also at the global level, the management strategies will be helpful to tackle the bad situation whenever it exists in near future. So, an integrated battle by the state, central global as well as stakeholders must be fought to win over the locust swarm and its upcoming bad impacts. Spraying pesticides with drones, aircraft, and other machinery do not provide a permanent solution. However, if the concept of integrated pest management (IPM) is adopted then it will benefit a lot and helps to tackle the situation conveniently with less effort. Besides the IPM approach, the basic cause of global warming should be known and then possible investments should be done in upgrading climatic conditions and new techniques, in order to save the farmers and other people from hunger.

KEYWORDS

- **locusts**
- **outbreaks**
- **Ladakh**
- **India**
- **plagues**
- **management**
- **losses**
- **economy**

REFERENCES

Abushama, F. T. Loss of Water from the Grasshopper *Poecilocerus hieroglyphicus* (Klug), Compared with the Tree Locust *Anacridium melanorhodon melanorhodon* (Walker). *Zeitschrift Angew. Entomol.* **1970,** *66*, 160–167.

Belayneh, Y. T. Acridid Pest Management in the Developing World: A Challenge to the Rural Population, a Dilemma to the International Community. *J. Orthopt. Res.* **2005,** *14*, 187–196.

Belovsky, G. E.; Slade, J. B. Insect Herbivory Accelerates Nutrient Cycling and increases plant production. *Proc. Natl. Acad. Sci. USA* **2000,** *97*, 14412–14417.

Brader, L.; Djibo, H.; Faye, F. G.; Ghaout, S.; Lazar, M.; Luzietoso, P. N. et al. *Towards a More Effective Response to Desert Locusts and Their Impacts on Food Security, Livelihoods and Poverty. Multilateral Evaluation of the 2003–05 Desert Locust Campaign.* Food and Agriculture Organisation: Rome, 2006.

Branson, D. H.; Joern, A.; Sword, G. A. Sustainable Management of Insect Herbivores in Grassland Ecosystems: New Perspectives in Grasshopper Control. *Bio Sci.* **2006,** *56*, 743–755.

Cigliano, M. M.; Braun, H.; Eades, D. C.; Otte, D. *Orthoptera Species File*. Version 5.0/5.0, 2017.

Cullen, D. A.; Cease, A. J.; Latchininsky, A. V.; Ayali, A.; Berry, K.; Buhl, J. et al. From Molecules to Management: Mechanisms and Consequences of Locust Phase Polyphenism. In *Advances in Insect Physiology*; Verlinden, H., Ed.; Elsevier: Oxford, 2017; pp 167–285.

D'Alessandro, S.; Fall, A. A.; Grey, G.; Simpkin, S.; Wane, A. *Senegal*; Agricultural Sector Risk Assessment: Washington, DC, 2015.

El-Mallakh, O. S.; El-Mallakh, R. S. Insects of the Qur'an. *Am. Entomol.* **1994,** *40*, 82–84.

Farrow, R. A. Population Dynamics of the Australian Plague Locust, *Chortoicetes terminifera* (Walker), in Central Western New South Wales. I. Reproduction and Migration in Relation to Weather. *Austr. J. Zool.* **1979,** *27*, 717–745.

Haroon, W. M.; Pages, C.; Vassal, J.-M.; Abdalla, A. M.; Luong-Skovmand, M.; Lecoq, M. Laboratory and Field Investigation of a Mixture of *Metarhizium acridum* and Neem Seed Oil Against the Tree Locust *Anacridium melanorhodon melanorhodon* (Orthoptera: Acrididae). *Biocontrol Sci. Technol.* **2011,** *21*, 353–366.

ICRISAT, Locust Swarm and Its Management. 29 May 2020.

Joern, A. Experimental Study of Avian Predation on Coexisting Grasshopper Populations (Orthoptera: Acrididae) in a Sandhills Grassland. *Oikos* **1986,** *46*, 243–249.

Latchininsky, A. V. Locusts and Remote Sensing: A Review. *JARS* **2013,** *7*, 075099.

Lockwood, J. A.; Showler, A. T.; Latchininsky, A. V. Can We Make Locust and Grasshopper Management Sustainable? *J. Orthoptera Res.* **2001,** *10*, 315–329.

Mohamed S. I. An overview on disasters. *Disaster Prev. Manag.* **2007,** *16*, 687–703.

Nevo, D. The Desert Locust, *Schistocerca gregaria*, and Its Control in the Land of Israel and the Near East in Antiquity, with Some Reflections on Its Appearance in Israel in Modern Times. *Phytoparasitica* **1996,** *24*, 7–32.

Oedekoven, M. A.; Joern, A. Plant Quality and Spider Predation Affects Grasshoppers (Acrididae): Food-Quality-Dependent Compensatory Mortality. *Ecology* **2000,** *81*, 66–77.

Osimani, A.; Garofalo, C.; Aquilanti, L.; Milanović, V.; Cardinali, F.; Taccari, M. et al. Transferable Antibiotic Resistances in Marketed Edible Grasshoppers (*Locusta migratoria migratorioides*). *J. Food Sci.* **2017,** *82*, 1184–1192.

Pener Locust Phase Polymorphism and Its Endocrine Relations. In *Advances in Insect Physiology*; Evans, P. D., Ed.; Elsevier: London, 1991; pp 1–79.

Phiriyangkul, P.; Srinroch, C.; Srisomsap, C.; Chokchaichamnankit, D.; Punyarit, P. Effect of Food Thermal Processing on Allergenicity Proteins in Bombay Locust (*Patanga succincta*). *Int. J. Food Eng.* **2015,** *1*, 23–28.

Qing, Y. Q. L. Z. L.; Chunxian, Y. G. F. C. J. Feeding Habits and Economic Threshold of *Locusta migratoria tibetensis*. *Acta Phytophylac. Sin.* **2008,** *5*.

Roffey, J. *Locusts and Grasshoppers of Economic Importance in Thailand*; Locusts and grasshoppers of economic importance in Thailand: London, 1979.

Schmitz, O. J. Effects of predator hunting mode on grassland ecosystem function. *Science* 319, 952–954.

Sokol-Hessner, L.; Schmitz, O. J. Aggregate Effects of Multiple Predator Species on a Shared Prey. *Ecology* **2002,** *83*, 2367–2372.

Steedman, A. *Locust Handbook*; Natural Resources Institute: Chatham, 1990.

Stige, L. C.; Chan, K.-S.; Zhang, Z.; Frank, D.; Stenseth, N. C. Thousand-Year-Long Chinese Time Series Reveals Climatic Forcing of Decadal Locust Dynamics. *Proc. Natl. Acad. Sci. USA* **2007,** *104*, 16188–16193.

Stoops, J.; Crauwels, S.; Waud, M.; Claes, J.; Lievens, B.; Van Campenhout, L. Microbial Community Assessment of Mealworm Larvae (*Tenebrio molitor*) and Grasshoppers (*Locusta migratoria migratorioides*) Sold for Human Consumption. *Food Microbiol.* **2016,** *53*, 122–127.

Tian, H.; Stige, L. C.; Cazelles, B.; Kausrud, K. L.; Svarverud, R.; Stenseth, N. C. et al. Reconstruction of a 1,910-y-Long Locust Series Reveals Consistent Associations with Climate Fluctuations in China. *Proc. Natl. Acad. Sci. USA* **2011,** *108*, 14521–14526.

Wiens, J. A. Pattern and Process in Grassland Bird Communities. *Ecol. Monogr.* **1973,** *43*, 237–270.

CHAPTER 10

Locusts Are Food for the Future to Supplement Various Nutritional Needs and Tackling Food Security for Saving Economic Budgets: It Is Hard to Defeat Them, Why Not to Eat Them?

BARKAT HUSSAIN[1], SHAFIA HASSAN[2], SALIQA SALWAT[1], RAYEES AHMAD[2], and TARIQ AHMAD[2]

[1]*Division of Entomology, SKUAST-K, Srinagar, Jammu & Kashmir, India*

[2]*Department of Zoology, University of Kashmir, Srinagar, Jammu & Kashmir, India*

ABSTRACT

Locusts are migratory pests that often take aerial routes, in the form of massive and massive swarms, and are used as food in many forms. If we recall history, they have been used as food since centuries, well documented in the literature of various types. Various cultures and countries consume locusts, especially in Africa, Middle East, and Asian countries. Locusts are used as feed and food for animals and poultry birds. They are vital sources of proteins. The idea of consuming them is a good idea as food supplements as animal feed and for human population is of no exception. They move from one country to another country in huge numbers and these massive swarms could be easily trapped and exploited as food

Locust Outbreaks: Management and the World Economy. Umair Riaz, Khalid Rehman Hakeem, (Eds.)

for the future to ensure food security crisis in the poorest of the poor countries. Though such cultures allow them to eat locusts. These locusts were important diets for North Americans. A single adult of locust consumes two grams of fresh vegetation per day. The solitary and gregarious swarms of locusts are threat to food security and livelihoods. Besides, to bring down these swarms under control, millions of dollars are being allocated by the world body (FAO) to save hunger of the world's poorest countries and at least 20 million people are at risk and finally to prevent the 25 million ha of cultivated areas are under threat. Locust consumption as food prevents malnutrition and protein scarcity in human population and also food and feed to other domesticated animals and poultry birds, just to prevent other food grains to feed them. The consumption of locusts for human population as food and other animals will ensure food security and thus can be exploited for the management of locusts in different parts of the globe to some extent. The exploitation of locusts as food will save different economic budgets allocated for the control of locust management, to address protein deficiencies, malnutrition in poorest of the poor countries (very low-income countries), potential crop damage, and to ensure a clean environment in the globe. The food saved in this direction is a great service to humanity as the sense of understanding is to knocking out the world's budget allocated for the management of locusts will ensure a food for thought for replicating our future policies and strategies at the global level.

10.1 INTRODUCTION

When compared to 2012, the food demand for 10 billion populations is expected to grow by around 50% in 2050 (FAO, 2017). Currently, malnourished people are around 820 million, two billion are deficient in micronutrients, and obese are two billion (Willett et al., 2019). Diet-related obesity and non-communicable diseases such as coronary heart disease, stroke, and diabetes are on the rise as a result of malnutrition (Willett et al., 2019). Traditional food diets are gradually not sustainable in terms of the environment, as they place strain on land and water resources, and positive impact on greenhouse gas emissions. Animal feed contributing to 60–80% costs to aquaculture, pigs, and poultry. The protein content of animal feed contributing to 70% costs (Dobermann et al., 2017; Mariod,

2020). Human population plagued by locust plagues for millennia. Locusts are members of the Acrididae family, and belong to the order Orthoptera, representing 6787 species (Cullen et al., 2017). Locusts, irrespective of grasshoppers have a cryptic solitarious phase, to a swarming gregarious phase. They migrate massively. The solitary locusts provide food for a variety of animals. Gregarious voracious locust swarms migrate for a variety of reasons. The fundamentals which trigger locust swarming are due to favorable environmental circumstances, such as global warming and human activity (Le Gall et al., 2019; Peng et al., 2020). Desert locust plagued in the Middle East, East Africa, and Asia during 2019–2020. The swarms spread to 2400 km^2. Such types of swarms were considered threat to 20 million populations and may lead to starvation (Roussi, 2020). Spraying aerially highly toxic insecticides are still a typical practice in locust control around the globe (Roussi, 2020), and are harmful on environment and human population (Van der Valk, 200). The potential for using locusts for decent purposes is really a justifiable management method. Desert locusts contain a high concentration of proteins, lipids, vitamins energy, and minerals (Kinyuru, 2020). They consume phytosterols from their diet and metabolize it into new compounds, which have health benefits for humans (Cheseto et al., 2015).

10.2 HISTORICAL BACKGROUND

Syrian and Assyrian monarchs in the second and first millennia BC, reported the capture of desert locusts for eating, and their slaughtering to avoid catastrophic crop losses (Lanfranchi, 2005). Locust eating was an aristocratic practice: a Neo Assyrian relief on a palace slab from the eighth century BC depicts the introduction of sets of locusts placed on spits to a royal meal. Queen Ranavalona II (1829–1883) of Madagascar directed attendants to harvest the locust (Locusta migratoria capito) for food (Decary, 1937). Mechanical and cultural management (trampling beating, pasture burning, and excavating up egg beds), used to suppress locusts, as noted by Price and Brown (2000) for the brown locust, and digging trenches, beating, and burning for the desert locust (Dobson, 2001). There was quite armory of human and horse-drawn machinery to smash out the Rocky Mountain locust, Melanoplus spretus, and leads to its extinction in the 1890s (Lockwood, 2004: 49–56).

10.3 DISTRIBUTION OF LOCUSTS

Twenty-one locust species are found across the globe (Cullen et al., 2017; Le Gall et al., 2019). Out of these 21 species of locusts, ten species are being consumed on animals for their feeding, in 65 countries. The migratory, desert, red, brown, Sahelian, Bombay, Australian plague, Italian, Siberian, and South American locusts are quite common (Yhoung-Aree, 2010; Kelemu et al., 2015; Costa-Neto and Dunkel, 2016; Mitsuhashi, 2016; Poma et al., 2017). Asian locust and European locusts are two subspecies of the migratory locusts. The northern Eurasian steppe is home to Italian, Siberian, and migratory locusts. While as, the Siberian locust is being found in the Sahara Desert and southern Europe. The Australian and African migratory locusts are native to Australia. African migratory locust is being found in Mainland Africa, Madagascar, Tibetan plateau and also in Southeast Asia. Bombay locust is found in India and also in Southeast Asia. The Sahelian locust is a native of North and Central Africa and also found in the Middle East. The red locust can be found in the wild.

10.4 IMPORTANCE OF LOCUSTS

The capacity of locust species is to migrate long distances and grow their population rapidly. The desert locust is the most significant locust species. In Africa, the Middle East, and south–west Asia, desert locust plagues can affect up to 20% (29 million km^2) of the earth's surface. Other important locust species include the desert locust (Locusta migratoria migratorioides), the migratory locust (Locusta migratoria migratorioides), the red locust (Nomadacris septemfasciata), and the brown locust (Nomadacris septemfasciata) in Africa (Locustana pardalina). The Australian plague locust, Chortoicetes terminifera, is found in Australia; the South American locust, Schistocerca cancellate, is found in South America; and the Oriental migratory locust, Locusta migratoria manilensis, and the Bombay locust, Nomadacris succincta, are found in Asia. The desert locust and the Italian locust, Calliptamus italicus (Western Europe to Central Asia), the Moroccan locust, Dociostaurus maroccanus (north–west Africa to Asia), and tree locusts, Anacridium spp., are found on multiple continents (Africa, Mediterranean, Near East).

10.5 COUNTRIES USING LOCUSTS AS FOOD

For millennia, 65 countries are harvesting locusts widely and devoured during epidemics. The countries like Africa and Asia (Mitsuhashi, 2016; van Huis, 2020; Weingarten, 2005) are the main countries. In different countries/customs, they are used for eating in a variety of methods, like boiling, toasting, roasting, frying, sun-drying, or even raw. Their flavor is similar to walnuts and almonds (Weingarten, 2005). In 26 African countries, five locust species are consumed, such as migratory, desert, red, brown, and Sahelian locusts (Kelemu et al., 2015; Mitsuhashi, 2016). Mexico, Kuwait, and Saudi Arabia all eat red locust (Mitsuhashi, 2016). In Australia, plague locusts are consumed. China people are consuming Italian and Siberian locusts. While in Brazil, South American locusts are being eaten. Adult migrating locusts are eaten in Madagascar (Van Itterbeeck et al., 2019). They are sometimes harvested by drawing large sheets over the plants during upsurges (such as the one from 2014 to 2016). Only winged adults are caught, boiled, and dried in the sun, frequently on highways. Pigs and poultry are fed with them. Attempts were made to harvest locusts in the Horn of Africa, the Arabian Peninsula, and the Indo–Pakistan border during the upsurge of 2019 to 2021 (FAO, 2021a). Concerned about pesticides and their environmental impact, an initiative in Pakistan pleased farmers of about US$ 0.125 each kg of locusts collected at night (Khan, 2020). Animal feed factories transform the insects into chicken feed (Samejo et al., 2021). Farmers were paid $ 1.65 per kg of collected locusts by the rural municipality. According to Reuters, a Kenyan organization was paying farmers US$ 0.45 per kg of locusts (Ratner, 2021).

10.6 NUTRITIONAL BENEFITS OF LOCUST

10.6.1 PROTEIN, FAT, AND ENERGY CONTENT

The protein, fat, and energy content of locusts are considered higher than meat (18–29%, 1–32%, and 106–353 kcal/100 g, respectively (Ahmad et al., 2018; Peng et al., 2020). Locusts contain high crude protein and chitin present in locusts increase solubility. Chitin of locusts is being extracted for other purposes (Shahidi et al., 1999). The higher chitin concentration is being used to estimate insect protein content of migratory locusts and other insects by 17%.

10.6.2 AMINO ACIDS

Despite this fact, the two amino acids (lysine and methionine) are found in fewer amounts in locusts. The recommended values are 4–6 times greater in locusts than in meat (Ahmad et al., 2018). In locusts, lysine presents in fewer amounts and methionine presents in larger amounts; whereas methionine is a protein source that is found in animal feed and also in soyabean meal (Makkar et al., 2014).

10.6.3 MINERALS

Desert locusts contain 1–6 and 2.4–12.5 mg/100 g of calcium (Ahmad et al., 2018). It has a higher level of micronutrients like iron and zinc than animal meat (Ahmad et al., 2018; Sun-Waterhouse et al., 2016). Zinc levels in migrating locusts are again higher than in animal meat. While heavy metal range is within acceptable limits in the locust than fruits and vegetables (Ahmad et al., 2018; Poma et al., 2017).

10.7 HEALTH BENEFITS OF LOCUSTS

Important sterols (sitosterol, campesterol, and stigmasterol) present in desert locusts, aid in the prevention of high blood pressure and heart disease (Cheseto et al., 2015; Sun-Waterhouse et al., 2016). The angiotensin-converting enzyme transforms inactive angiotensin I into octapeptide angiotensin II in desert locusts (Vercruysse et al., 2005). They can be utilized as antihypertensive compounds in nutraceuticals. The angiotensin-converting enzyme, ACE inhibitors, viz; captopril, has adverse side effects. Cooked and baked desert locusts when hydrolyzed, multiply skin fibroblasts in humans, important for the formation of extracellular matrix components in the skin (Zielinska et al., 2015). Outward indications of aging are caused by advancing aging, which restricts fibroblast proliferation. Finally, consuming processed locusts can help to minimize the aging symptoms.

10.8 LOCUSTS AS FEED

In fish feed, desert and migratory locusts can replace up to 25% of the dietary protein (Makkar et al., 2014). However, substituting locust meal for more than 25% of the time decreases fish performance, because of high chitin, that hinders the protein solubility (Brogan, 2018). The addition of undefined quantity of red locust into pig diets caused acceptable development (Makkar et al., 2014). From a red locust-based diet, from which bacon and pork is being generated, had a fishy flavor, which could be avoided by stopping the diet 3 weeks before slaughter. Arginine, a necessary amino acid for laying chickens, is said to be lower in locusts than in soybeans (Makkar et al., 2014). However, locust-based meals can be supplemented with low-cost synthetic amino acids (Makkar et al., 2014).

10.9 HARVESTING, PROCESSING, AND MARKETING OF LOCUSTS

10.9.1 HARVESTING OF LOCUSTS

The majority of the locust swarms are diurnal. Harvesting of locusts can be done in large numbers during night (Raheem et al., 2019; van Huis, 2020). Thai government launched operation to stimulate the collection of Bombay locusts for human consumption when all control tactics totally failed (Hanboonsong (2010), to suppress their infestation. The Bombay locust is now one of the most common edible insects in Thailand. Few farmers in Thailand are now cultivating maize for locust feeding (Hanboonsong, 2010; Hanboonsong et al., 2013). In East Africa, desert locust pandemic, a Kenyan company, track and capture locusts for processing into protein-rich powder for animal composition (Driver, 2021). The Bug Picture used 180 harvesters during the course of a 2-month pilot to collect 4.3 tonnes of locusts and earn $1900. In Pakistan, a similar approach was used to harvest roughly 3 tonnes of desert locust (Samejo et al., 2021). Locust hoppers can also be mass-harvested (Sharma, 2014; Mitsuhashi, 2016), by pushing them into long trunches. Recent methods for collecting locust crowds are with broom sweeping/nets, manual collection, and also in trenches (Samejo et al., 2021; van Huis, 2020). Furthermore, in Russia, light-emitting diode lamps were efficient in harvesting swarms of unspecified locust species (Lysakov et al. (2019). In New South Wales, 14,000 solitary Australian plague locusts were trapped per night, using 150 W UV lamps (Farrow (1974). Direct light repelled the Australian plague locusts. In East Africa, there is a big chance to build effective commercial locust light traps (Sengendo et al., 2021) and also for edible grasshoppers.

10.9.2 PROCESSING

Apart from sun drying, the common methods of locust processing are fermenting with salt, roasting, boiling, and frying (Mariod, 2020; Mutungi et al., 2019; Weingarten, 2005). These procedures have a significant impact on the nutritional and functional qualities of locusts. Protein from boiled locust samples showed the highest digestibility of 50% as compared to 41% in fried locust samples in the Sudan market (Mutungi et al., 2019). Processing methods increase mineral levels and water and fat absorption capability. The value chain for locust consumption necessitates for developing procedures on nutritional profiles.

10.9.3 MARKETING OF LOCUSTS

The edible insect market was estimated to be worth US$33 million in 2015 in the United States, Belgium, France, the United Kingdom, Vietnam, Brazil, and Mexico. The edible insect market in South Korea had been estimated as US $142 million during 2017 (Dobermann et al., 2017). Trade of edible insects worth $1.14 million each year, including locusts, had been (Hanboonsong et al., 2013) reported in Thailand. Locusts are highly valued by Thai consumers, costing US$7–11/kg (Hanboonsong et al., 2013). While as in some countries, live and cooked locusts are imported from various market channels, especially in the off-season (Grabowski et al., 2020; Yhoung-Aree, 2010). The world edible insect market is expected to surpass US$1.5 billion by 2026, with no distinction made between wild-caught and farmed insects.

10.10 MAJOR HEALTH AND ENVIRONMENTAL SAFETY ISSUES ASSOCIATED WITH LOCUSTS

To have efficient monitoring of locusts, their biology and ecology is very good for any control programs. The application of insecticides is still the weapons for the control of locusts in foreseeable future. However, biopesticides are being used to control locusts, and satellite imagery and geographic information (Zhang et al., 2019) do play a key role to target the locusts effectively and timely, when they become airborne. In addition to predict the future outbreaks and their range expansion and their sizes, the preventive solutions are highly real, more viable, and sound but require additional technology advancement before being implemented on a broad scale (Zhang et al., 2019). Aerial application of insecticides is still the most trustworthy technique for managing the locusts in the event of a plague (Roussi, 2020). Of course, toxic chemical residues can accumulate in locusts, posturing serious health issues to humans and animals (Makkar et al., 2014; Mariod, 2020; Mutungi et al., 2019). In Arabia and Kuwait, organophosphate residues were found in locusts (Van der Spiegel et al., 2013). Spraying insecticides on locusts had undesirable impact on non-target invertebrates and other aquatic species, and also mass deaths of freshwater shrimp. Therefore, safer research alternatives could be explored for using locusts for food and feed. The easiest method is mass trapping of unsprayed locusts to fetch a good price and their safe use for consumption purposes.

10.11 SOCIAL AND CULTURAL ISSUES ASSOCIATED WITH LOCUST CONSUMPTION

Locust consumption is influenced by cultural traditions and socioeconomic variables (Sun-Waterhouse et al., 2016). The Holy books, the Quran and the Bible, forbid consuming any insects except locusts, and few insects. Locusts are healthy food in Africa and Asia. The Europeans consume standard food which should be blended with locust flour (Mitsuhashi, 2016; Sun-Waterhouse et al., 2016). Young European males and females, practice entomophagy, while Chinese and Germans also consume locusts.

10.12 RULES AND REGULATIONS FOR LOCUST CONSUMPTION

Globally, national food safety agencies have recently started, to review edible insect rules and guidelines. European Union-approved insect proteins in fish feed (Costa-Neto and Dunkel, 2016) and yellow mealworm larva in human food (IPIFF, 2021). The production, consumption, and sale of migratory locusts are permitted in the Netherlands and Belgium. While, Switzerland and Canada permitted their use as food and feed for human population (Halloran et al., 2015; Sun-Waterhouse et al., 2016). In Thailand, products of edible insects, must be accepted by the Food and Drug Administration for marketing purposes. The use of insects as feed ingredients has been permitted in Kenya and Uganda (IDRC, 2019).

10.13 CONCLUSION

Locusts are nutritionally equivalent to, if not superior to, traditional meat. They are low in cholesterol and contain omega-3 and 6 fatty acids, as well as other sterols, which are important in preventing heart disease. The limiting amino acids lysine and methionine are 4–6 times higher in locusts than they are in regular beef. Locusts are high in calcium, iron, zinc, vitamin D3, B12, E, and A, as well as heavy metals in tolerable amounts. Efforts are underway around the world to build regulatory frameworks and laws that favor locust cultivation, consumption, and commerce. We must advocate establishing value chains to harness locust swarms for positive applications by raising public awareness about their nutritional worth, addressing safety concerns, establishing regulations, and developing technologies for efficient collecting and processing. The consumption of locusts as food and feed purposes not only saves trillion liters of water, pesticides, and other resources for their management. The use of locusts as food and feed boost the economy and leads to the concept of entomophagy in the real world and economic boost and boom at global level.

KEYWORDS

- **locust**
- **food and feed**
- **economic budgets**
- **policies**
- **strategies**
- **food security**
- **malnutrition**

REFERENCES

Ahmad, R. S.; Imran, A.; Hussain, M. B. Nutritional Composition of Meat. In *Meat Science and Nutrition*; Arshad, M. S., Ed.; IntechOpen: London, 2018; pp 61–77.

Brogan E. N. Protein and Lipid Characterization of Acheta domesticus, Bombyx mori, and Locusta migratoria Dry Flours. MSc, West Virginia University, USA, 2018.

Cheseto, X.; Kuate, S. P.; Tchouassi, D. P.; Ndung'u, M.; Teal, P. E.; Torto, B. Potential of the Desert Locust Schistocerca gregaria (Orthoptera: Acrididae) as an Unconventional Source of Dietary and Therapeutic Sterols. *PloS One* 2015, *10*, e0127171. https://doi.org/10.1371/journal.pone.0127171

Cullen, D. A.; Cease, A. J.; Latchininsky, A. V.; Ayali, A.; Berry, K.; Buhl, J.; De Keyser, R.; Foquet, B.; Hadrich, J. C.; Matheson, T.; Ott, S. R.; Poot-Pech, M. A.; Robinson, B. E.; Smith, J. M.; Song, H.; Sword, G. A.; Vanden Broek, J.; Verdonk, R.; Verlinden, H.; Rogers, S. M. From Molecules to Management: Mechanisms and Consequences of Locust Phase Polyphenism. In *Advances in Insect Physiology*; Verlinden, H., Ed.; 2017; pp 167–285. https://doi.org/10.1016/bs.aiip.2017.06.002

Costa-Neto, E. M.; Dunkel, F. Insects as Food: History, Culture, and Modern Use Around the World. In *Insects as Sustainable Food Ingredients: Production, Processing and Food Applications*; Dossey, A. T., Morales-Ramos, J. A., Rojas, M. G., Eds.; Academic Press: Cambridge, 2016; pp 29–60. https://doi.org/10.1016/B978-0-12-802856-8.00002-8

Cheseto, X.; Baleba, S.; Tanga, C. M.; Kelemu, S.; Torto, B. Chemistry and Sensory Characterization of a Bakery Product Prepared with Oils from African Edible Insects. *Foods* 2020, *9*, 800. https://doi.org/10.3390/foods9060800.

Decary, R. L'entomophagie chez les indigènes de Madagascar. *Bulletin de la Société Entomologique de France. Juin* 1937, *9*, 168–171.

Dobermann, D.; Swift, J.; Field, L. Opportunities and Hurdles of Edible Insects for Food and Feed. *Nutr. Bull.* 2017, *42*, 293–308. https://doi.org/10.1111/nbu.12291.

Dong, Y. Transforming Insect Biomass into Consumer Wellness Foods: A Review. *Food Res. Int.* 2016, *89*, 129–151. https://doi.org/10.1016/j.foodres.2016.10.001

Driver, E. From Plague to Promise: How Insects Can Revolutionize Our Food Systems, 2021. https://grist.org/fix/from-plague-to-promise-how-insects-can-revolutionize-our-food-systems/.

Dobson, H. M. Desert Locust Guidelines: Control. *J. Insects Food Feed* 2001, *7* (3), 247.

FAO. The Future of Food and Agriculture—Trends and Challenges. Food and Agriculture Organization of the United Nations, Rome, 2017. http://www.fao.org/3/i6583e/i6583e.pdf

FAO. Desert Locust Information Service of the Migratory Pests Group: Frequently Asked Questions. Food and Agriculture Organization of the United Nations: Rome, 2020. http://www.fao.org/ag/locusts/oldsite/LOCFAQ.htm#q2.

FAO/WHO/UNU. Protein and Amino Acid Requirements in Human Nutrition, 2007.

Food and Agriculture Organization of the United Nations (FAO), 2014. Evaluation of Field Trials Data on the Efficacy and Selectivity of Insecticides on Locusts and Grasshoppers. Report to FAO by the Pesticide Referee Group 10th Meeting. December 10–12, 2014. Gammarth, Tunisia, 66 pp. http://www.fao.org/3/bu337e/bu337e.pdf.

Food and Agriculture Organization of the United Nations (FAO). Current Upsurge (2019–2021). Locust Watch, FAO, Rome, Italy, 2021a. http://www.fao.org/ag/locusts/en/info/2094/index.html

Food and Agriculture Organization of the United Nations (FAO). Frequently Asked Questions (FAQs) About Locusts. Locust watch, FAO, Rome, Italy, 2021b. http://www.fao.org/ag/locusts/en/info/info/faq/index.html.

Food and Agriculture Organization of the United Nations, Rome, Italy, 2001. http://www.fao.org/ag/locusts/common/ecg/347_en_DLG4e.pdf.

Gatheru, J. W.; Khamis, F. M.; Ombura, F. L. O.; Nonoh, J.; Tanga, C. M.; Maina, J.; Mohamed, S. A.; Subramanian, S.; Ekesi, S.; Fiaboe, K. K. M. Impact of Processing Methods on Microbial Load of Reared and Wild-Caught Edible Crickets (Scapsipedus icipe and Gryllus bimaculatus) in Kenya. *J. Insects Food Feed* 2019, *5*, 171–183. https://doi.org/10.3920/JIFF2018.0042

Grabowski, N. T.; Tchibozo, S.; Abdulmawjood, A.; Acheuk, F.; M'Saad Guerfali, M.; Sayed, W. A.; Plotz, M. Edible Insects in Africa in Terms of Food, Wildlife Resource, and Pest Management Legislation. *Foods* 2020, *9*, 502. https://doi.org/10.3390/foods9040502.

Halloran, A.; Vantomme, P.; Hanboonsong, Y.; Ekesi, S. Regulating Edible Insects: The Challenge of Addressing Food Security, Nature Conservation, and the Erosion of Traditional Food Culture. *Food Secur*. 2015, *7*, 739–746. https://doi.org/10.1007/s12571-015-0463-8.

Hanboonsong, Y. Edible Insects and Associated Food Habits in Thailand. In *Forest Insects as Food: Humans Bite Back, Proceedings of a Workshop on Asia-Pacific Resources and Their Potential for Development*. FAO Regional Office for Asia and the Pacific: Bangkok, 2010; pp 173–182.

Hanboonsong, Y.; Jamjanya, T.; Durst, P. B. *Six-legged Livestock: Edible Insect Farming, Collection and Marketing in Thailand*; Food and Agriculture Organization of the United Nations: Bangkok, 2013.

IDRC. Integrating Insects in Poultry and Fish Feed in Kenya and Uganda, 2019. https://ww.idrc.ca/en/research-in-action/integrating-insects-poultry-and-fish-feeds-kenya-and-uganda

IPIFF. *IPIFF Welcomes EU Member States' Green Light to the Authorisation of the First Insect Food Products at European Level*; The International Platform of Insects for Food and Feed: Brussels, 2021. https://ipiff.org/wp-content/uploads/2021/05/May-04-2021-PR-novel-food-vote-2-1.pdf

KEBS. Edible Insects' Products—Specification DKS 2922-1:2020: ICS 67.120. Kenya Bureau of Standards: Nairobi, 2020. https://www.kebs.org/images/standards/public_review_standards/2020/june/dks_2922part_1_2020_edible_insects_products_pr.pdf

Kelemu, S.; Niassy, S.; Torto, B.; Fiaboe, K.; Affognon, H.; Tonnang, H.; Maniania, N.; Ekesi, S. African Edible Insects for Food and Feed: Inventory, Diversity, Commonalities and Contribution to Food Security. *J. Insects Food Feed* 2015, *1*, 103–119. https://doi.org/10.3920/JIFF2014.0016.

Lanfranchi, G. B. Minilivestock Consumption in the Ancient Near East: The Case of Locust. In *Ecological Implications of Minilivestock: Potential of Insects, Rodents, Frogs and Snails*; Paoletti, M. G., Ed., Science Publishers, Inc.: Enfield, NH, 2005; pp 163–174.

Latchininsky, A.; Sword, G.; Sergeev, M.; Cigliano, M. M.; Lecoq, M. Locusts and Grasshoppers: Behavior, Ecology, and Biogeography. *Psyche* 2011, 578327. https://doi.org/10.1155/2011/578327.

Le Gall, M.; Overson, R. and Cease, A. A Global Review on Locusts (Orthoptera: Acrididae) and Their Interactions with Livestock Grazing Practices. *Front. Ecol. Evol.* 2019, *7*, 263. https:// doi.org/10.3389/fevo.2019.00263

Lockwood, J. A. *Locust: The Devastating Rise and Disappearance of the Insect That Shaped the American Frontier*; Basic Books: New York, 2004.

Lysakov, A.; Grinchenko, V.; Molchanov, A.; Devederkin, I. Effect of Ultra-Bright Led Light for Locust Plague Control. *Eng. Rural Dev*. May 22–24, Jelgava, Latvia, 2019; pp 630–634. https://doi.org/10.22616/erdev2019.18.n353

Lecoq, M.; Pierozzi Jr.; I. Rhammatocerus Schistocercoides Locust Outbreaks in Mato Grosso (Brazil): A Long-Standing Phenomenon. *Int. J. Sustain. Dev. World Ecol*. 1995, *2*, 45–53.

Makkar, H. P.; Tran, G.; Heuz´e, V.; Ankers, P. State-of-the-Art on Use of Insects as Animal Feed. *Anim. Feed Sci. Technol*. 2014, *197*, 1–33. https://doi.org/10.1016/j.Anifeedsci.2014.07.008.

Mariod, A. A. Nutrient Composition of Desert Locust (Schistocerca gregaria). In *African Edible Insects as Alternative Source of Food, Oil, Protein and Bioactive Components*; Mariod, A. A., Ed.; Springer: Cham, 2020; pp 257–263. https://doi.org/10.1007/978-3-030-32952-5_18.

Mitsuhashi, J. *Edible Insects of the World*; CRC Press: FL, 2016; p 274.

Musundire, R.; Osuga, I. M.; Cheseto, X.; Irungu, J.; Torto, B. Aflatoxin Contamination Detected in Nutrient and Anti-Oxidant Rich Edible Stink Bug Stored in Recycled Grain Containers. *PloS One* 2016, *11*, e0145914. https://doi.org/10.1371/journal.Pone.0145914.

Mutungi, C.; Irungu, F.; Nduko, J.; Mutua, F.; Affognon, H.; Nakimbugwe, D.; Ekesi, S.; Fiaboe, K. Postharvest Processes of Edible Insects in Africa: A Review of Processing Methods, and the Implications for Nutrition, Safety and New Products Development. *Crit. Rev. Food Sci. Nutr.* 2019, *59*, 276–298. https://doi.org/10.1080/10408398.2017.1365330

Paul, A.; Frederich, M.; Uyttenbroeck, R.; Hatt, S.; Malik, P.; Lebecque, S.; Hamaidia, M.; Miazek, K.; Goffin, D.; Willems, L. Grasshoppers as a Food Source? A Review. *Biotechnol. Agron. Soc. Environ*. 2016, *20* (S1), 337–352.

Price, R. E.; Brown, H. D. A Century of Locust Control in South Africa. In Workshop on Research Priorities for Migrant Pests of Agriculture in Southern Africa. Plant Protection Research Institute, Pretoria, South Africa, 24–26 March 1999. https://gala.gre.ac.uk/id/eprint/12211/1/12211_Cheke_Workshop%20on%20research%20priorities%20(book)%202000.pdf#page=51.

Ratner, B. Farmers Fight Back: Making Animal Feed from a Locust Plague. https://widerimage.reuters.com/story/farmersfight-back-making-animal-feed-from-a-locust-plague.

Roussi, A. The battle to contain gigantic locust swarms. Nature 579, 330.

Saeed, T.; Dagga, F. A.; Saraf, M. Analysis of Residual Pesticides Present in Edible Locusts Captured in Kuwait. *Arab Gulf J. Sci. Res*. 1993, *11*, 1–5.

Samejo, A. A.; Sultana, R.; Kumar, S.; Soomro, S. Could Entomophagy Be an Effective Mitigation Measure in Desert Locust Management? *Agronomy* 2021, *11*, 455. https://doi.org/10.3390/agronomy11030455

Sengendo, F.; Subramanian, S.; Kidoido, M.; Chemurot, M.; Tanga, C.; Egonyu, J. P. Cost–Benefit Analysis of Improved Light Trap for Harvesting the Edible Grasshopper, Ruspolia Differens (Orthoptera: Tettigoniidae): Evidence from Uganda, 2021.

Shahidi, F.; Arachchi, J. K. V.; Jeon, Y.-J. Food Applications of Chitin and Chitosans. *Trends Food Sci. Technol.* 1999, *10*, 37–51.

Sharma, A. Locust Control Management: Moving from Traditional to New Technologies—An Empirical Analysis. *Entomol.; Ornithol. Herpetol.* 2014, *4*, 2161–2183.

Van Huis, A. Insects as Food and Feed, a New Emerging Agricultural Sector: A Review. *J. Insects Food Feed* 2020, *6*, 27–44. https://doi.org/10.3920/JIFF2019.0017

Vercruysse, L.; Smagghe, G.; Herregods, G.; Van Camp, J. ACE Inhibitory Activity in Enzymatic Hydrolysates of Insect Protein. *J. Agric. Food Chem.* 2005, *53*, 5207–5211. https://doi.org/10.1021/jf050337q.

Weingarten, S. Wild Foods in the Talmud: The Influence of Religious Restrictions on Consumption. In *Wild Foods: Proceedings of the Oxford Symposium on Food and Cookery*; Hoskings, R., Ed.; Oxford Brookes University: Oxford, 2005; p 321.

Willett, W.; Rockstrom, J.; Loken, B.; Springmann, M.; Lang, T.; Vermeulen, S.; Garnett, T.; Tilman, D.; DeClerck, F.; Wood, A.; Jonell, M. Food in the Anthropocene: The EAT–Lancet Commission on Healthy Diets from Sustainable Food Systems. *Lancet* 2019, *393*, 447–492. https://doi.org/10.1016/S0140-6736(18)31788-4.

Xinhua. Nepali Municipal Governments, Farmers Make Efforts to Fight Locust Invasion. Xinhuanet, 14 July 2020. http:// www.xinhuanet.com/english/2020-07/14/c_139212263.htm.

Yhoung-Aree, J. Edible Insects in Thailand: Nutritional Values and Health Concerns. In *Forest Insects as Food: Humans Bite Back*; Durst, P. B., Johnson, D. V., Leslie, R. N., Shono, K., Eds.; Food and Agriculture Organization: Bangkok, 2010; pp 201–216.

Zhang, Z. S.; Lu, X. G.; Wang, Q. C.; Zheng, D. M. Mercury, Cadmium and Lead Biogeochemistry in the Soil–Plant–Insect System in Huludao City. *Bull. Environ. Contam. Toxicol.* 2009, *83*, 255–259. https://doi.org/10.1007/s00128-009-9688-6.

Zhang, L.; Lecoq, M.; Latchininsky, A.; Hunter, D. Locust and Grasshopper Management. *Annu. Rev. Entomol.* 2019, *64*, 5–34. https://doi.org/10.1146/annurev-ento-011118-112500.

Zielinska, E.; Baraniak, B.; Kara´s, M.; Rybczynska, K.; Jakubczyk, A. Selected Species of Edible Insects as a Source of Nutrient Composition. *Food Res. Int.* 2015, *77*, 460–466. https://doi.org/10.1016/j.foodres.2015.09.008

CHAPTER 11

Role of Pheromones in Aggregation and Reproduction in Locusts

NAJEEBUL TARFEEN[1], KHAIR-UL-NISA[2], ANJUM SABBA[3], ZAHRA SULTAN[4], and SABA WANI[3]

[1]Centre of Research for Development (CORD), University of Kashmir, Srinagar, India

[2]Department of Environmental Science, University of Kashmir, Srinagar, India

[3]Department of Biochemistry, University of Kashmir, Srinagar, India

[4]Department of Botany, University of Kashmir, Srinagar, India

ABSTRACT

Pheromones are the chemical messengers that have a profound impact on behavior as well as on physiology of other individuals of the same species. In Locust, *Schistocerca gregaria*, many volatiles have been reported to have a role in the gregarious nature and also in the reproductive behavior of individuals. Aggregated oviposition in females is as a result of pheromone-based behavior in interrelationship with environmental factors. Various studies displayed the role of pheromones in the timing of sexual maturation. Mature males secrete many volatiles, which facilitated the sexual maturity in immature males. The gregarious nature in locusts is actually a response to mechanical stimuli. Moreover, main constituent of

Locust Outbreaks: Management and the World Economy. Umair Riaz, Khalid Rehman Hakeem, (Eds.)

aggregation pheromone system in adults is phenylacetonitrile, is seen to have repellent properties and is not known to be associated with the aggregation. Pheromones secreted by gregarious males are known to intensify male guarding behavior upon inhibition of courtship hormone. In this chapter, various studies and perspectives regarding the role of pheromones in aggregation and reproduction in locusts is discussed.

11.1 INTRODUCTION

Locusts are capable of transforming reversibly between two main phases, one being solitaria and the other is gregaria. These two phases vary in their physiology, morphology, and behavior (Uvarov, 1966). Locusts have important feature of aggregating in their gregarious-phase. They form dense groups by clumping together, and swarm over great distances as adults (Steedman, 1988). Various factors have shown to influence the phase dynamics, which include visual, tactile, and chemical stimuli, besides that, the early history of locusts and dietary factors are also shown to affect phase characteristics of locusts (Obeng-Ofori et al., 1994b). It has been suggested that there are a variety of chemicals known as pheromones that have an involvement in the aggregation and reproduction (BYERS, 1991). The pheromones that have a part in the aggregation of locusts were first notified by (Nolte, 1963) and (Gillett, 1968, 1975). Aggregation mediated by pheromones helps insects to bring together for feeding, reproduction, and protection (Borden, 1985). Pheromones may be releaser pheromones that bring behavioral responses and other being primer pheromones that have a role in development, physiology as well as behavior. In male adults (desert), a number of volatiles has been noted such as benzaldehyde, phenylacetonitrile, veratrole, anisole, phenol, and guaiacol, and are mainly liberated by mature adults, especially by gregarious males (Luber et al., 1993). Torto et al. (1996) reported that in gregarious adult males, Phenylacetonitrile is the main volatile in the aggregation pheromone system and also a pronounced pheromone in causing maturation (Mahamat et al., 1993). In a study (Seidelmann and Ferenz, 2002), it found that PAN also acts as a repellent pheromone and does not prove to have role in aggregation rather has an involvement in sexual behavior. Investigations by (Guo et al., 2020) reported that 4-vinylanisole has a part in aggregation in *Locusta migratoria.* It was reported that some volatiles in *L. migratoria manilensis* were sex specific, that is, male specific volatiles only affects adult males while as female specific volatiles affects only females and has no effect on males (Li and Zhang, 2018). This chapter gives information that how sex volatiles affect the courtship and mating behavior in adult forms and also gives an account of aggregation pheromone system and its impact on the gregarization of Locusts.

11.2 AGGREGATION PHEROMONE SYSTEM

Locusts are capable of transforming reversibly between two extreme phases, solitary and gregarious. In gregarious locusts, there occurs a change in behavior and tends to aggregate, they begin to clump together and swarm over great distances. Aggregation behavior and phase transition phenomenon are as a result of pheromonal release. While analyzing the volatile compounds, results in affirmation of the releaser pheromone, being peculiar to nymphs and primer pheromone being distinctive to adult ones (Obeng-Ofori et al., 1993; Torto et al., 1996). Aggregation is a elaborative event facilitated by events like attraction as well as cohesion stimulated by pheromones (Fuzeau-Braesch et al., 1988). Nolte (1963) in his experiment found that when third-instar "clumped" hoppers were segregated and transferred to another room, they lost their pigmentation gradually at each molt and changed to sandy-colored suggested the presence of pheromone in the surrounding air of clumped hoppers. The pheromone stimulated the melanization, is a characteristic of the gregarious phase. Gillett (1968) found the evidence for the pheromonal role in aggregation in desert locusts, causing solitary locusts to show clumping behavior after exposure to air from crowded locusts and lose such aggregation behavior if raised in clean air. Nolte et al. (1970) introduced the term gregarization pheromone to include above-described effects. In a study by Gillett (1975) found another evidence of aggregation pheromone, where she observed that solitary hoppers of desert locusts raised in clean air, often undergo an extra instar when compared to locusts exposed to air containing pheromones from clumped locusts. The time of development to attain the fifth instar stage was less for pheromone-exposed ones (21 days) in comparison to solitary hoppers in clean air (37 days). In a study on *L. migratoria,* it was observed that pheromones from hoppers lessens the time period of fifth instar and thus synchronize moulting (Nolte, 1976).

In adult locusts (desert), a number of volatiles have been reported such as, benzaldehyde, phenylacetonitrile, veratrole, anisole, phenol, and guaiacol, and most of them are released by mature adults, particularly by gregarious males (Luber et al., 1993). Olfactometer bioassays were used to study the aggregation pheromone system in locusts and led to the conclusion that aggregation pheromone is specific to adult male locusts (Torto et al., 1994). A study by Guo et al., 2020 reported that 4-vinylanisole acts as an aggregation pheromone in *Locusta migratoria*. Reports revealed that both solitary and gregarious ones are attracted toward it regardless of their age and sex.

11.3 SEX PHEROMONES

Finding a mate and reproduction requires diverse strategies in solitaries and gregarious locusts. To attract a male in less dense population over large distances is a key for successful reproduction in desert locusts. Various studies revealed that volatiles secreted by female may increase the chances of sexual meeting in a solitary field (Inayatullah et al., 1994). One such volatile compound secreted by female is E,Z-2,6 nonodienal which induces a significant EAG response in males. Visual stimuli were found to be important for female findings in contrast to acoustic signals that do not proved to be essential for male response.

In desert locusts, mating would take place only when developed eggs are carried by females and when they are about to oviposition. It is at this stage that acceptance of males by females is high for courtship. During that period, behavioral changes occur in males by secretion of volatiles by females. For instance, pentanoic acid secreted by females during oogenesis process may act as releaser pheromones thereby activating the premating behavior in male adults (Njagi and Torto, 2002). A study by (Li and Zhang, 2018) suggested that gregarious locusts employs certain volatiles that governs their biological attributes and many of these chemicals are sex-specific. Using GC–MS, six volatiles that were sex specific were found in fecal samples of *L. migratoria manilensis,* among them, male-specific volatile (2-heptanone) was found to be repellent to adult males and having no effect on females. Alternately, female specific (2,2,6-trimethylcyclohexanone) was repugnant to other adult females and was having no impact on males.

11.3.1 COURTSHIP INHIBITION BEHAVIOR OF LOCUSTS

The mating behavior in *S. gregaria* has been studied in detail by (Popov, 1958) and (Uvarov, 1977). The mating involves nearing of male to female discontinued by breaks in a semi-stilted position. But overall no such perplexed course of courtship behavior has been noticed in locusts

(Popov, 1958). In densely populated conditions, main problem for male *S. gregaria* will the sperm competition instead of mate finding. The male that copulates last during courtship with female prior to oviposition will be the father of the majority of offsprings (Hunter-Jones, 1960). Thus *S. gregaria* has to protect its inseminated female to fend off other males so to prevent sperm competition until those sperms are being utilized to fecundate the eggs. The male stick over their mate and shields her till oviposition occurs. In addition to that, in a swarm where there occurs high competition for females, males guard the females by male concealment strategies (Simmons and Siva-Jothy, 1998). Gregarious males of sexually mature desert locusts release a considerable quantity of phenylacetonitrile (PAN) (Seidelmann et al., 2000). This pheromone is liberated only when mature males are present in vicinity and the quantity released is dependent on the sum total of rivals present. It is suggested that PAN acts as courtship inhibition pheromone, by concealing the responsive female and shielding her from other rival males (Seidelmann and Ferenz, 2002). PAN acts as some type of olfactory camouflage that seems to mask the receptivity-signaling female and shields the pair from other competitive males. It has a function similar to antiaphrodisiac to prevent males from attempts to mount the female body. This pheromone seems to be efficacious for mate guarding. Moreover, this pheromone also acts as a signal supplemental to gregarious yellow coloration thereby labeling mature males. Phenylacetonitrile averts males from useless homosexuality. This type of pheromone is known as abstinence (Schlein et al., 1981).

In actual sense, Phenylacetonitrile is as a repellant pheromone (Seidelmann et al., 2000). Same reports were suggested by (Ignell et al., 2001) that PAN is repugnant to other males. Some studies revealed that it acts as an aggregation pheromone (Torto et al., 1994). To conclude these studies, response to phenylacetonitrile is age dependent and juvenile hormone too has a role. Moreover, it is observed that PAN is linked with both sexual maturities of males as well as sexual competition indicating its role in reproductive behavior in high density population. In migratory locusts, it was observed that extended precopulatory mounting enhances the time period of mating leading to high proportion of offspring impregnated by second male to copulate (Zhu and Tanaka, 2002). In *S. gregaria*, no mating plug is laid down by male that might obstruct the female reproductive tract, unlike found in African migratory locust (*L. migratoria)*. In *L. migratoria*, mate guarding behavior prior to copulation has been studied by Parker and Smith, 1975 and males are known to plug spermatophore by clogging up spermathecal duct of female till oviposition. In *L. migratoria*, no need to protect the female after copulation to prevent from sperm competition. The *S. gregaria* has to shield the impregnated mate to fend off competitors and prevents sperm competition. Moreover, in desert locusts copulation would take place when a female holds developed eggs and is about to oviposit. It is only this time period female accepts courting male.

11.4 ROLE OF MATURATION ACCELERATING PHEROMONE IN DESERT LOCUSTS

It is suggested that pheromones impart a role in the timing of sexual maturation in desert locusts. The two-set of primer pheromones has been observed to regulate the maturation in desert locusts, one being maturation-accelerating pheromone related with adult males and other being maturation retarding pheromone linked with immature instars of insect (Norris, 1954) (Richards and El Mangoury, 1968) (Amerasinghe, 1978). These pheromones assists in simultaneous mating and oviposition (Norris and Richards, 1963). Studies revealed an accelerating impact of volatiles by adult males on sexual maturation of juveniles that includes both females and males. These volatiles are secreted by epidermal cells and are oil soluble in nature (Loher, 1961). Various other volatiles were analyzed and detected by GC–MS, among them benzaldehyde, veratrole, 4-vinylveratrole, anisole, and phenylacetonitrile were seen to have accelerating effect on sexual maturation (Mahamat et al., 2000). In various experiments, immature males were kept in contact with mature ones without having any visual or physical contact or were exposed to synthetic mixtures of pheromone compounds. Both synthetic combinations of volatiles as well as presence of mature males enhanced the appearance of bright coloration (yellowish) in males and also affects their sexual behavior. The exclusion of anisole from synthetic mixture had no significant effect. But exclusion of any of the other analytes slowed down the sexual maturation substantially. PAN in particular, appears to be crucial for accelerating sexual maturity. However, investigations done by Schmidt and Albütz, (2002) suggested that neither PAN solely nor other volatiles accelerates the sexual maturity of juvenile males, but pentane compound extracted from cuticular surface of yellow mature males has the capability to enhance sexual maturity. Assad et al. (1997) have reported the decelerating impact of nymphal volatiles (unidentified) on the sexual maturity of immature adult males. Experimental studies demonstrated that males that are sexually mature accelerate the vitellogenin levels in females. Both approaches, whether slowing down of maturation by nymphal volatiles or speeding up of maturation by secretions from mature males are supposed to concur with the development of gregarious locusts (Mahamat et al., 2000). Pener (1991) reported that sexual maturity in *S. gregaria* is controlled by juvenile hormone (JH). Activation of endocrine system by these volatiles may lead to the modification of juvenile hormone secretion by corpora allata, maturation of gonads, and finally affect the mating behavior.

11.5 BIOSYNTHESIS OF PHEROMONES THAT HAVE ROLE IN AGGREGATION, SEXUAL MATURATION, AND REPRODUCTION

Pheromones in females that are linked with oviposition as well as reproductive behavior, seems to originate from Comstock-Kellog glands, that is, female accessory glands (Njagi and Torto, 2002). Pheromones of aggregation pheromone system such as phenol and guaiacol are linked with the feces. It was postulated that these volatiles were liberated by gut bacteria, as in other locusts that were germ-free, no guaiacol and very little amounts of phenol have been reported (Obeng-Ofori et al., 1994a; Dillon et al., 2000). Studies demonstrated that differences between mature and juvenile locusts could be attributed to the food quality and lesser count of gut bacteria (Dillon and Charnley, 2002). From these results it can be concluded that these phase-related discrepancies in the quantitative and qualitative constituents of volatiles may be because of variations in food quality and metabolism. Additionally, it should be specified that not every volatile released by locusts and inducing an EAG response has specific chemical information for locusts.

Phenylacetonitrile, was supposed to be produced by abdominal epidermal cells (Luber, 1994). These cells are large in gregarious phase of adult males, but are underdeveloped in solitary as well as in juvenile females and males of *S. gregaria*. Besides, their presence in abdominal cells, reports are there that reveal its presence in legs (tarsal segments) (Kendall, 1972). Investigations done by (Seidelmann et al., 2003) suggested that legs as well as wings are the sites of production of phenylacetonitrile. When tissues of legs and wings were incubated using C^{14}-phenylalanine as precursor leading to the liberation of C^{14}-PAN confirming its production from wings and legs. (Amwayi et al., 2012) reported that PAN although found in both sexes, is released only by sexually mature gregarious male adults. This study also revealed that external tissues of both sexes have greater distribution of the pheromone as compared to internal tissues. Another study reported that this pheromone was obtained from faecal extracts of mature/older males (Obeng-Ofori et al., 1994a). Veratrole, another volatile compound is known to be secreted by all parts in an adult male locust (Seidelmann et al., 2003).

Phenylacetonitrile biosynthesis seems to be under the regulation of neuroendocrine system. In a study done by (Ferenz and Seidelmann, 2003) reported that methanolic extracts of brains from mature male desert locusts, accelerated the release of PAN when injected into gregarious mature males that were previously put into isolation, suggesting that neuropeptide regulates the synthesis of the pheromone. Secretion of PAN is limited to sexually adult males; juvenile hormone has a part in inducing the development of those tissues that produce pheromones or assist in the release of potential biosynthesis-activating neurohormone. Tawfik et al., 2000 reported the association of JH-titre with sexual maturation and pheromonal release. Application of JH III affected the pheromonal production (delayed production) in young adult males of locusts (Tawfik et al., 1997).

11.6 CONCLUSIONS

Studies have suggested that a large number of volatiles have been identified that are linked with locusts. The aggregation behavior in locusts is a multifactorial complex involving the phase shifting from solitary to gregarious ones. Stimulus in the form of tactile, olfactory, and visual cues is responsible for specific alterations of some of the phase traits. Locusts at different stages release diverse types of odors that do not prove that these secretions are pheromones. Responses to these secretions/volatiles can be stage or phase-specific and JH has a role in modifying the responsiveness. The role of phenylacetonitrile as a repellent to competing males, are emphasizing the significance of strategies in mating among crowded locust populations. However, much remains to be done on the role of these volatiles in the solitary phase, comprehensive study of reproductive diapause and mating in solitary forms in connection with vegetation patterns as well as weather conditions would provide a foundation for developing population and aggregation models and for forecasting locust outbreaks.

KEYWORDS

- **methanolic extracts**
- **sex phermones**
- **chemical stimulation**
- **reproduction**
- **phenylacetonitril**
- **larva**

REFERENCES

Abdellaoui, K., et al. The Aggregation Pheromone Phenylacetonitrile: Joint Action with the Entomopathogenic Fungus *Metarhizium anisopliae* var. Acridum and Physiological and Transcriptomic Effects on *Schistocerca gregaria* Nymphs. *Pesticide Biochem. Physiol.* **2020,** *167,* 104594.

Aly, S.A.; ElEbiarie, A.S.; Hamadah, K.S. Effects of the Wild Plant, *Fagonia bruguieri* on the Adult Performance and Phase Transition of *Schistocerca gregaria* (Orthoptera: Acrididae). *Egypt. Acad. J. Biol. Sci. A, Entomol.* **2010,** *3*(2), 133–147.

Amerasinghe, F. Pheromonal Effects on Sexual Maturation, Yellowing, and the Vibration Reaction in Immature Male Desert Locusts (Schistocerca gregaria). *J. Insect Physiol.* **1978,** *24* (4), 309–314.

Amwayi, P. W.; Masiga, D. K.; Govender, P.; Teal, P. E.; Torto, B. Mass Spectral Determination of Phenylacetonitrile (PAN) Levels in Body Tissues of Adult Desert Locust, Schistocerca gregaria. *J. Insect Physiol., 58* (8), 1037–1041.

Assad, Y. O.; Hassanali, A.; Torto, B.; Mahamat, H.; Bashir, N. H.; El Bashir, S. Effects of Fifth-Instar Volatiles on Sexual Maturation of Adult Desert Locust Schistocerca gregaria. *J. Chem. Ecol.* **1997,** *23* (5), 1373–1388.

Borden, J. *In Comprehensive Insect Physiology, Biochemistry and Pharmacology*, Vol. 9. Edited by Kerkut, G. A., Gilbert, L. I.; Pergamon Press: Oxford, 1985.

BYERS, J. A. Pheromones and Chemical Ecology of Locusts. *Biol. Rev.* **1991,** *66* (4), 347–378.

Dillon, R.; Charnley, K. Mutualism Between the Desert Locust Schistocerca gregaria and Its Gut Microbiota. *Res. Microbiol.* **2002,** *153* (8), 503–509.

Dillon, R. J.; Vennard, C. T.; Charnley, A. K. Exploitation of Gut Bacteria in the Locust. *Nature* **2000,** *403* (6772), 851–851.

Ferenz, H. J.; Seidelmann, K. Pheromones in Relation to Aggregation and Reproduction in Desert Locusts. *Physiol. Entomol.* **2003,** *28* (1), 11–18.

Fuzeau-Braesch, S.; Genin, E.; Jullien, R.; Knowles, E.; Papin, C. Composition and Role of Volatile Substances in Atmosphere Surrounding Two Gregarious Locusts, Locusta migratoria and Schistocerca gregaria. *J. Chem. Ecol.* **1988,** *14* (3), 1023–1033.

Gillett, S. Airborne Factor Affecting the Grouping Behaviour of Locusts. *Nat., 218* (5143), 782–783.

Gillett, S. D. Changes in the Social Behaviour of the Desert Locust, Schistocerca gregaria, in Response to the Gregarizing Pheromone. *Anim. Behav.* **1975,** *23*, 494–503.

Guo, X.; Yu, Q.; Chen, D.; Wei, J.; Yang, P.; Yu, J. et al. 4-Vinylanisole Is an Aggregation Pheromone in Locusts. *Nature* **2020,** *584* (7822), 584–588.

Hunter-Jones, P. Fertilization of Eggs of the Desert Locust by Spermatozoa from Successive Copulations. *Nature* **1960,** *185* (4709), 336–336.

Ignell, R.; Couillaud, F.; Anton, S. Juvenile-Hormone-Mediated Plasticity of Aggregation Behaviour and Olfactory Processing in Adult Desert Locusts. *J. Exp. Biol.* **2001,** *204* (2), 249–259.

Inayatullah, C.; El Bashir, S.; Hassanali, A. Sexual Behavior and Communication in the Desert Locust, Schistocerca gregaria (Orthoptera: Acrididae): Sex Pheromone in Solitaria. *Environ. Entomol.* **1994,** *23* (6), 1544–1551.

Kendall, M. D. Glandular Epidermis on the Tarsi of the Desert Locust, Schistocerca gregaria Forskal. *Acrida*, 1972.

Kumari, A. and N. Kaushik, *Oviposition deterrents in herbivorous insects and their potential use in integrated pest management*. 2016.

Latchininsky, A.V., *Locusts and remote sensing: a review.* Journal of Applied Remote Sensing, **2013.** 7(1): p. 075099-075099.

Li, J.; Zhang, L. Two Sex-Specific Volatile Compounds Have Sex-Specific Repulsion Effects on Adult Locusts Locusta migratoria manilensis (Meyen)(Orthoptera: Acrididae). *Entomol. News* **2018,** *127* (4), 293–302.

Li, J. and L. Zhang, T*wo Sex-Specific Volatile Compounds Have Sex-Specific Repulsion Effects on Adult Locusts Locusta migratoria manilensis (Meyen)(Orthoptera: Acrididae)*. Entomological News, **2018.** *127*(4): p. 293–302.

Loher, W. The Chemical Acceleration of the Maturation Process and Its Hormonal Control in the Male of the Desert Locust. *Proc. R. Soc. London B Biol. Sci.* 1961, *153* (952), 380–397.

Luber, K. *Untersuchungen zur chemischen Okologie der Wustenheuschrecke Schistocerca gregaria*; PhD Thesis, University of Oldenburg, 1994.

Luber, K.; Wieting, J.; Zeeck, E.; Ferenz, H. Isolation and Characterization of a Volatile Aromatic Infochemical Released by Sexual Maturing Gregarious Male Desert Locusts, Schistocerca gregaria. *Verhandlungen der Deutschen Zoologischen Gesellschaft* **1993,** *86*, 261.

Mahamat, H.; Hassanali, A.; Odongo, H. The Role of Different Components of the Pheromone Emission of Mature Males of the Desert Locust, Schistocerca gregaria (Forskål)(Orthoptera: Acrididae) in Accelerating Maturation of Immature Adults. *Int. J. Trop. Insect Sci.* **2000,** *20* (1), 1–5.

Mahamat, H.; Hassanali, A.; Odongo, H.; Torto, B.; El-Bashir, E.-S. Studies on the Maturation-Accelerating Pheromone of the Desert Locust Schistocerca gregaria (Orthoptera: Acrididae). *Chemoecology* **1993,** *4* (3), 159–164.

Nishide, Y., T. Suzuki, and S. Tanaka, *Synchrony in the hatching of eggs in the desert locust Schistocerca gregaria (Orthoptera: Acrididae): egg condition influences hatching time in the laboratory and under simulated field temperatures.* Applied entomology and zoology, **2017.** *52:* p. 599–604.

Njagi, P. G.; Torto, B. Evidence for a Compound in Comstock-Kellog Glands Modulating Premating Behavior in Male Desert Locust, Schistocerca gregaria. *J. Chem. Ecol.* **2002,** *28* (5), 1065–1074.

Nolte, D. A Pheromone for Melanization of Locusts. *Nature* 1963, *200* (4907), 660–661.

Nolte, D. Locustol and Its Analogues. *J. Insect Physiol.* **1976,** *22* (6), 833–838.

Nolte, D.; May, I.; Thomas, B. The Gregarisation Pheromone of Locusts. *Chromosoma* **1970,** *29* (4), 462–473.

Norris, M. J. Sexual Maturation in the Desert Locust with Special Reference to the Effects of Grouping. *Anti-Locust Bull.* **1954,** *18*, 1–43.

Norris, M. J.; Richards, M. O. Laboratory Experiments on Gregarious Behaviour in Ovipositing Females of the Desert Locust,(Schistocerca gregaria (forsk.)). *Entomologia experimentalis et applicata* **1963,** *6* (4), 279–303.

Obeng-Ofori, D.; Torto, B.; Hassanali, A. Evidence for Mediation of Two Releaser Pheromones in the Aggregation Behavior of the Gregarious Desert Locust, Schistocerca gregaria (Forskal)(Orthoptera: Acrididae). *J. Chem. Ecol.* **1993,** *19* (8), 1665–1676.

Obeng-Ofori, D.; Torto, B.; Njagi, P. G.; Hassanali, A.; Amiani, H. Fecal Volatiles as Part of the Aggregation Pheromone Complex of the Desert Locust, Schistocerca gregaria (Forskal) (Orthoptera: Acrididae). *J. Chem. Ecol.* **1994a,** *20* (8), 2077–2087.

Obeng-Ofori, D.; Njagi, P.; Torto, B.; Hassanali, A.; Amiani, H. Sex Differentiation Studies Relating to Releaser Aggregation Pheromones of the Desert Locust, Schistocerca gregaria. *Entomologia experimentalis et applicata* **1994b,** *73* (1), 85–91.

Parker, G.; Smith, J. Sperm Competition and the Evolution of the Precopulatory Passive Phase Behaviour in Locusta Migratoria Migratorioides. *J. Entomol. A Gen. Entomol.* **1975,** *49* (2), 155–171.

Pener, M. Locust Phase Polymorphism and Its Endocrine Relations. *Adv.Insect Physiol.* 1991, *23*, 1–79.

Popov, G. B. Ecological Studies on Oviposition by Swarms of the Desert Locust (Schistocerca gregaria Forskal) in Eastern Africa. *Ecological Studies on Oviposition by Swarms of the Desert Locust (Schistocerca gregaria Forskal) in eastern Africa* (31), 1958.

Richards, M.; El Mangoury, M. Further Experiments on the Effects of Social Factors on the Rate of Sexual Maturation in the Desert Locust. *Nature* **1968,** *219* (5156), 865–866.

Rizvi, S.A.H., et al., *Latest developments in insect sex pheromone research and its application in agricultural pest management.* Insects, **2021.** *12*(6): p. 484.

Rono, E., et al., *Concentration-dependent parsimonious releaser roles of gregarious male pheromone of the desert locust, Schistocerca gregaria.* Journal of insect physiology, **2008.** *54*(1): p. 162–168.

Schlein, Y.; Galun, R.; Ben-Eliahu, M. Abstinons. *J. Chem. Ecol.* 1981, *7* (2), 285–290.

Schmidt, G. H.; Albütz, R. Sexual Maturation and Yellow Coloration of Adult Males in the Gregarious Desert Locust, Schistocerca gregaria, in Relation to Volatiles Emitted (Caelifera: Acrididae). *Entomologia generalis* **2002,** 121–141.

Seidelmann, K.; Ferenz, H.-J. Courtship Inhibition Pheromone in Desert Locusts, Schistocerca gregaria. *J. Insect Physiol* **2002,** *48* (11), 991–996.

Seidelmann, K.; Luber, K.; Ferenz, H.-J. Analysis of Release and Role of Benzyl Cyanide in Male Desert Locusts, Schistocerca gregaria. *J. Chem. Ecol.* **2000,** *26* (8), 1897–1910.

Seidelmann, K.; Weinert, H.; Ferenz, H.-J. Wings and Legs Are Production Sites for the Desert Locust Courtship-Inhibition Pheromone, Phenylacetonitrile. *J. Insect Physiol.* **2003,** *49* (12), 1125–1133.

Simmons, L.; Siva-Jothy, M. 10 Sperm Competition in Insects: Mechanisms and the Potential. *Sperm Competition Sexual Selection* 1998, 341.

Sorensen, P.W. and C. Baker, *Species-specific pheromones and their roles in shoaling, migration, and reproduction: a critical review and synthesis.* Fish pheromones and related cues, **2014**: p. 11–32.

Steedman, A. *Locust Handbook*, 1988.

Tawfik, A. I.; Osir, E. O.; Hassanali, A.; Ismail, S. H. Effects of Juvenile Hormone Treatment on Phase Changes and Pheromone Production in the Desert Locust, Schistocerca gregaria (Forskal)(Orthoptera: Acrididae). *J. Insect Physiol.* **1997,** *43* (12), 1177–1182.

Tawfik, A.I. and F. Sehnal, *A role for ecdysteroids in the phase polymorphism of the desert locust*. Physiological entomology, **2003.** *28*(1): p. 19–24.

Tawfik, A. I.; Treiblmayr, K.; Hassanali, A.; Osir, E. Time-Course Haemolymph Juvenile Hormone Titres in Solitarious and Gregarious Adults of Schistocerca gregaria, and Their Relation to Pheromone Emission, CA Volumetric Changes and Oocyte Growth. *J. Insect Physiol.* **2000,** *46* (7), 1143–1150.

Torto, B.; Njagi, P. G.; Hassanali, A.; Amiani, H. Aggregation Pheromone System of Nymphal Gregarious Desert Locust, Schistocerca gregaria (Forskål). *J. Chem. Ecol.* **1996,** *22* (12), 2273–2281.

Torto, B.; Obeng-Ofori, D.; Njagi, P. G.; Hassanali, A.; Amiani, H. Aggregation Pheromone System of Adult Gregarious Desert Locust Schistocerca gregaria (Forskal). *J. Chem. Ecol.* **1994,** *20* (7), 1749–1762.

Torto, B., et al., *Odor composition of field versus laboratory desert locust populations.* Journal of Insect Physiology, 2021. **134**: p. 104296.

Uvarov, S. B. Grasshoppers and Locusts. A Handbook of General Acridology. Volume I. Anatomy, Physiology, Development, Phase Polymorphism, Introduction to Taxonomy, 1966.

Uvarov, S. B. *Grasshoppers and Locusts. A Handbook of General Acridology, Vol. 2*; Behaviour, Ecology, Biogeography, Population Dynamics: Centre for Overseas Pest Research, 1977.

Wei, J., et al., *Composition and emission dynamics of migratory locust volatiles in response to changes in developmental stages and population density.* Insect science, **2017.** *24*(1): p. 60–72.

Wertheim, B., et al., *Pheromone-mediated aggregation in nonsocial arthropods: an evolutionary ecological perspective*. Annu. Rev. Entomol., **2005.** *50*: p. 321–346.

Zhu, D.-H.; Tanaka, S. Prolonged Precopulatory Mounting Increases The Length of Copulation and Sperm Precedence in Locusta migratoria (Orthoptera: Acrididae). *Ann. Entomol. Soc. Am.* **2002,** *95* (3), 370–373.

CHAPTER 12

Influence of the COVID-19 Pandemic and Locust Outbreaks on the World's Economy

UMAIR RIAZ[1], KHALID REHMAN HAKEEM[2], HUMERA AZIZ[3], and SANA FAROOQ[4]

[1]*Department of Soil & Environmental Sciences, MNS University of Agriculture Multan, Pakistan*

[2]*Department of Biological Sciences, Faculty of Science, King Abdulaziz University, Jeddah, Saudi Arabia*

[3]*Department of Agricultural Sciences, Government College University, Faisalabad, Pakistan*

[4]*Department of Environmental Sciences, Government College University, Faisalabad, Pakistan*

ABSTRACT

The world has experienced a lot of change during the last two decades, economically and in terms of human health and food security. The locust swarm severely affected major croplands in Africa, Asia, and Australia while COVID-19 pandemic affected the world economically, financially, emotionally, and medically eventually dramatically. These events were pushed 50 million more families toward below the poverty line. But the

Locust Outbreaks: Management and the World Economy. Umair Riaz, Khalid Rehman Hakeem, (Eds.)

Locust swarm is happening almost now in every year which has the potential threat to affect agriculture badly and ultimately food security in all over the world. Each normal swarm could include as many as 1.5 billion grasshoppers each km^2 and, therefore, can travel up to 150 km in a single day. Just a little locust flock from 1 km^2 may consume the same quantity of food as 35,000 humans in a single day. These circumstances when joined with COVID-19, the damage to humans was beyond countries' boundaries and economy. It is the need of the time, to explain all these circumstances and events in detail for the better understanding of measures, controls, and management strategies to overcome farmers' losses and to ensure food security. This chapter covers the major losses of COVID-19 on agriculture and locust swarm in Asia, Africa, and Australia.

12.1 INTRODUCTION

The World's economy is affected by the COVID-19 epidemic. It disrupted different channels of countries' economies by slowing it down. As a result, country's provisional GDP growth for 2019–2020 is expected to be negative, particularly in the regions where Locust and COVID-19 are affected simultaneously. In Pakistan, locust swarms initially appeared in the provinces of Baluchistan and Khyber Pakhtunkhawa, followed by Sindh and southern Punjab. In Sindh province, the combined effect of COVID-19 outbreak and locust swarm creates serious worries regarding agriculture output and food safety. Although, the prolonged effects of COVID-19 epidemic and locust swarms will turn out to be more obvious in the next months. This concisely shows their underlying effects on ranch families in rustic Sindh. Agroindustry is fundamental to Pakistan's economy, offering more than 19.0% of GDP (PBS, 2020) and utilizing around 39% of the workforce (PBS, 2018). Farming is the strength of the provincial people, which establishes 63% of the nations' all-out populace (PBS, 2017) and supplies a huge portion of Pakistan's commodities (Economic Adviser's Wing, Ministry of Finance, 2020). A survey employing computer-assisted telephone interviewing was undertaken in June 2020 with the help of the Asian Development Bank (ADB).

1. This study aimed to reach 721 farmers throughout eight districts in Sindh Province, and 410 farmers were successfully interviewed. Here, the survey gathered data about how COVID-19-related restrictions or interruptions impacted both harvest and sale for

Rabi (winter) seasonal crops and dairy items, their availability and cost for supplies, but also farmers' income commitments.

2. To conclude the effect of grasshoppers attack in the locale, a survey was also solicited. This survey revealed that more than half (58.3%) of responders consume less food in their families. About a quarter (37.3%) of respondents said that their families experienced troubles in compensation and nonfarm income as a result of COVID-19. Also, one family member had gotten back from city and other regions stated by 39.5% respondents. According to 45.4% interviewees, ranch families are troubled by an expansion in the family members and decreased money pay, which result in lower nonfood consumption.

Producers of all types of crops (counting wheat, tomato, organic products, and vegetables) and dairy have been adversely influenced by coronavirus-related issues. More than 65% of wheat growers and over 67% of fruits and vegetables growers revealed hardships through marketing their crops. Tomato growers faced a significantly more extreme result; more than 32% couldn't market their crops by any means. Additionally, 61.2% had the option to market their crops, however, with trouble; also, 61% of respondents couldn't finish their tomato yield. Buyers became often unable or reluctant to buy milk from dairy farmers, which impacted milk farmers as well. A total of 81% of milk farmers said they haven't been able to sell their product on a daily basis in the last few months.

12.2 CORONAVIRUS AND RESTRICTIONS ON FOOD SUPPLY CHAINS

On 31 December 2019, WHO was informed of cases of pneumonia of unknown cause in Wuhan City, China. A novel coronavirus was identified as the cause by Chinese authorities on 7 January 2020 and was temporarily named "2019-nCoV." The first verified case of coronavirus in Pakistan was confirmed on 26 February 2020 in Sindh Province. COVID-19 instances had been filed in four states; the two independent regions; and the government capital, Islamabad. As reaction to COVID-19 pandemic epidemic, the Federal government implemented a half lockdown start from 22 March 2020 in all states except Sindh. In Sindh, the provincial government had forced a total lockdown in the first week of March because of the significant number of affected individuals. Despite the fact that the limits were only partially applied to the agriculture sector and the food supply chain in order to ensure enough food availability in local markets, disruptions such as higher rental fees, workers shortage, and growers' limited access to markets were noted (FAO, 2020a). Different reports claimed anecdotal proof of ranchers having significant difficulty getting toward selling their crops to market in April and May (Jamal, 2020; Action Against Hunger, 2020). In Pakistan generally, farmers sell their crops to discount markets (Siddiq and Basher, 2019), although studies show that some are unable to get their crops to market due to mobility issues (Latif and Niazi, 2020). Many non-agricultural economic activities were almost stopped as a result of the COVID-19-related countrywide lockdown, which might have a negative influence on food supply chains. To sustain the availability of food throughout seasons and to yield the benefit of varied agro-biological zones, Pakistan depends generally on interprovincial food mobility (FAO, 2020a). Wheat is a primary product related to food security through the interprovincial food supply. Another comparable phone survey taken in Punjab (ADB, 2020) indicated that the lockdown had a severe impact on the cultivation and sale of fresh foods such as fruits, vegetables, and milk, which could have problems for Pakistan's food security. COVID-19 had an influence on farm supplies as well, by 55.2% of respondents saying an increase in the price of their agricultural supplies as a result of the virus, which might disrupt harvest in the coming season and cause economic suffering for farm owners. The planting of kharif yields and collecting of rabi crops started soon after lockdown was announced. Sindh and Balochistan start their kharif growth phase earlier (in April) than Punjab and Khyber Pakhtunkhwa (in May). Ranchers who are growing crops faced more difficulties because their development phase coincided with the lockdown (third seven-day stretch of March–9 May). Consequently, Sindh ranchers who had early harvesting stage confronted main difficulties in collecting and selling their crops, compounded by work lack and transport issues (Khan et al., 2020). Within the short term, these restrictions can increase food security concerns (United Nations Development Programme, 2020).

12.3 LOCUST SWARMS AND FOOD SECURITY

The FAO noted "An upsurge [of locusts] developed in 2019 as a result of two cyclones that brought heavy rains to the Empty Quarter on the Arabian

Peninsula in May and October 2018 that was exacerbated by Desert Locust outbreaks along the Red Sea coast during the winter of 2018 and 2019. The cyclones allowed at least three generations of unprecedented breeding in the Empty Quarter that was not detected. Swarms emigrated from these areas to spring breeding areas in the Central and Eastern regions from January to March 2019. Two generations of spring breeding occurred that spread to the Horn of Africa and to the Indo–Pakistan border in June. Three more generations occurred in the latter area as a result of the best monsoon rains in 25 years, leading to locust attacks in June 2020." Seven Locust swarms struck Pakistan's provinces Khyber Pakhtunkhwa and Baluchistan first, then Sindh and southern Punjab. According to the FAO, locust attacks might cost up to Rs. 688.5 billion ($4.1 billion) in kharif crop loss and PRs705.8 billion ($4.2 billion) in rabi crop loss if no proper actions are taken, expecting 25% harm to crops (Dowlatchahi et al., 2020). The invasion of that level has not happened over the most recent 25 years, and the obsolete government framework for insect control is unprepared to battle the new grasshopper assaults. Swarms of locusts continue to be a problem. Grasshoppers are yet present in Baluchistan close to the border line of Sindh, noticed by FAO (footnote 5). They also forecast that beetles will fledge and shape grown-up gatherings and little crowds along the Indo–Pakistan line, which will develop, lay eggs, and cause a second (but more modest) rush of grasshoppers in October (FAO, 2020b). Previously in 2020, the central government stated a national locust emergency, permitting several government departments to work together in the fight. Although the locust danger to agriculture is not bigger than that presented by COVID-19, their emergence at the same time might have multiple effects with disastrous consequences. As a result, the study aimed to evaluate the effect of the locust attack, which has surely increased worries of food security, as well as whether the government had supplied ranchers with early notice of the swarms. The consolidated impacts of the continuous COVID-19 pandemic, the increase of desert locust, and ongoing extreme floods are posing an unexpected triple socio-economic danger to the East African area. Again, COVID-19 epidemic with the shocking financial effects of desert beetle and flooding brought about destroying effects on income and food security circumstances in East Africa. Overall main objective of this research was to examine as well as synthesize information from Ethiopia upon these extensive socio-economic effects of the three dangers in East Africa. For achieve this aim, all necessary data were gathered under this study using PubMed/PMC/Medline, Web of Science, Google Scholar, Scopus, Google, and Science Direct databases' internet sites. Local paper, news relevant to the topic from public entities, international organizations, and non-governmental organizations (NGOs) were also utilized. The survey tracked down that the dangers don't simply affect financial states of individuals in the locale overall and in Ethiopia specifically. However, they also worsen existing food insecurity and challenge livings and progress achievements that have taken years to achieve. Government-imposed restrictions aimed at containing the spread of the COVID-19 epidemic within their borders will have negative socioeconomic consequences for their citizens. Ethiopia's economy will suffer an irreversible loss as a result of the epidemic, which will be compounded by the devastating impacts of floods and desert locust. Its GDP is predicted to decline by 3.0% in 2019/2021. The conclusions of this review will assist and inform that the COVID-19 epidemic, desert swarmer, and floods were all witnessed as well as require governmental measures to prevent future repetition. It is critical for countries to be prepared for any of these risks in order to avoid negative socioeconomic consequences and human lives being lost. As a result, governments within every nation within area must make realistic promises to achieve socioeconomic development in an acceptable time frame. In particular, the Ethiopian government must implement policies and funds to strengthen people's risk tolerance ability to manage with and mitigate the negative consequences of these disasters within the immediate term, as well as to construct a stronger, comprehensive, and viable future on the long base. This worldwide COVID-19 epidemic is a lifetime experience disaster unlike anything else after World War II ended (Catherine, 2020). The COVID-19 is one of humankind's worst critical concerns in recent history, impacting every country as well as all sectors of society (WFP, 2020a). This epidemic too has left major imprints on the world economic and social economic balance, in addition to major health difficulties. It's true in the sense that financial effects coming about because of the COVID-19 emergency are relied upon to discriminate affect a wide scope of financial and social markers (Beyene et al., 2020). The current type of COVID-19 reaction had forced people who are already socially excluded in many ways into a precarious situation because of the absence of significant social government assistance and particularly pay support. The COVID-19 epidemic or its aftermath have wreaked havoc on the world's healthcare, social, and economic structures (Kimani et al.,

2020). As COVID-19 spreads, its condition might quickly deteriorate (Herring et al., 2020). COVID-19 epidemic and efforts to control it have had severe economic consequences, including a noteworthy decline in the world's economy (Josephson et al., 2020). Every research attempting to analyze the impact of the COVID-19 epidemic on income, poverty, or other welfare and medical variables must depend on macroeconomic situations that analyze projected paths of macroeconomic factors in a uniform manner (Hurnik et al., 2020). Other than the direct financial effects of the well-being crisis, the epidemic will harm poor nation by worldwide supply chain disturbances, the tourism industry failure, products value falls, falls in settlements by transients, capital flights, and overseas speculation decay (Raga, 2020). Furthermore, East Africa recently suffered one of the heaviest rainfall spells in 40 years. Resulting in sever wide flooding, dislocating a huge number of individuals and causing crop (70,000 hectares) and animals harm (96,000 animals passing) in the most exceedingly terribly damaged regions. The excessive rains have helped the reproduction and growth of Desert Locusts (DL) in East Africa, extending the outburst that remains to represent a critical danger to the main agricultural seasons. Because of its capacity to breed quickly, move vast areas, and ruin crops, the DL is ranked a most destructive of any migratory insect species in the world (Cressman, 2016). Each normal swarm could include as many as 1.5 billion grasshoppers each square kilometer and, therefore, can travel up to 150 kilometers in a single day. Just a little locust flock from one square kilometer may consume the same quantity of food as 35,000 humans in a single day (Food and Agricultural Organization [FAO], 2020d). The circumstance when joined with COVID-19, floods and the spread of these insects created "Triple Menace for East African individuals as per (Global Agricultural Monitoring [GAM], 2020). COVID-19, DL, and dangerous flooding represent a "Triple Threat" to a large number of individuals across East Africa.

That additional threat of COVID-19 disclosure is a sub-area that previously was home-based to around 20% of the total populace of food-shaky individuals. Such crises are much more than just brief consequences; also increase existing food shortages and threaten incomes and development achievements, which have taken years to achieve (WFP, 2020b). Because of sub-region is primarily made up of low-income nations, the consequences of these tragedies are severe. Kenya, South Sudan, areas of Eastern and Southern Ethiopia, and Somalia continue stretched to disaster levels of food anxiety from December 2019 to January 2020 as per research to (Famine Early Warning Systems Network [FEWS NET], 2020a). The effect of the DL just on sub-food area's and livelihood security is deeply damaging. Eastern Africa's present DL behavior is particularly concerning (International Federation of Red Cross and Red Crescent Societies [IFRC], 2020a). The FAO has warned that the DL epidemic is endangering food production in East Africa, citing considerable grassland and crop losses, primarily in different agricultural areas of Eastern Ethiopia, Central Somalia, and Northern Kenya (FAO, 2020a, 2020c). Due to the combined effects of COVID-19, DL, and floods, the World Food Programme estimates that the number of food that people in a sub-region might rise to 34 million or even more over the next months. Furthermore, as per the study of Integrated Food Security Phase Classification [IPC] (2020a), the area faces boundless food uncertainty, with around 12.7 million individuals in Crisis (IPC Phase 3) or more in Ethiopia, Kenya, and Somalia (8.5 million in Ethiopia, 3.1 million in Kenya, and 1.1 million individuals in Somalia). COVID-19, climatic shocks (unusual floods), terrorism, DL, and macroeconomic problems are all factors that contribute to food poverty (IPC, 2020a). This spread of both the COVID-19 pandemic or the locust-endemic disasters might lead to starvation, health-hazards, and poverty in the affected areas (Pescaroli & Alexander, 2018). According to existing data, nearly 130 million of people may risk famine by its end of 2020 as a result of COVID-19's estimated effect on the economy, most often in locust-infested low- and middle-income African nations (Rahaman et al., 2020). Additional concerns posed both COVID-19 and the locust infestation are being worsened by the current flood situation. Flooding is a "Threat Amplifier" when it comes toward the spreading of COVID-19 because it makes it more difficult to establish preventative measures in place (IFRC, 2020b). Flood events are often the most common and deadly natural calamity on the planet, resulting in massive fatalities, illnesses, and infrastructure and agricultural loss. Because of a lack of adequate disaster management systems and financial means, floods do have a larger influence on poor nations (Abaya et al., 2008). Because of the spatially compounded impact of the COVID-19 epidemic, DL, plus unprecedented floods, the existing social economic condition of East Africa is severely worrying. It poses an unexpected danger for people's macroeconomic conditions, particularly their food shortages and livelihood possibilities inside the country. Throughout Ethiopia, the continent's second-largest country after Nigeria, the effect of such dangers is particularly damaging and severe, as even the country has been weakened by the cumulative effects of political destabilization, conflict, ethnic tensions, drought, malnutrition, and economic instability throughout its history. As a result, such variables have increased the macroeconomic effects of these risks, severely reducing the country's threat-resistance capability.

Hence, these factors have aggravated the socioeconomic impacts of these threats and highly deteriorated the risk resilience capacity of the people in the country. It meant that there has been insufficient proof of the risks' severe effects on the well-being of people in East Africa in general, and Ethiopians in especially, for making short- and long-term arrangements and precautions. However, whereas studies on COVID-19's health effects in low-income countries are increasing fastly (Mbow et al., 2020), there is insufficient data on the epidemic economic effects (Laborde et al., 2020). As a result, this evaluation has become important, but it can also help governments and foreign assistance organizations (stakeholders) make decisions on how to effectively reduce the triple threats' adverse social economic effects and brought them to light in East Africa.

Considering the general condition in East Africa, as well as the influence of DL on food security, the recent locust population that started in July 2019 had been epidemic in Africa, and this is a mishap with a cascading risk when combined with the COVID-19 outbreak (Pescaroli & Alexander, 2018). The DL epidemic in East Africa is affecting lives, food production, ecology, and regional economic growth (Abubakr et al., 2020). In Ethiopia and Somalia, the present rise is even worse in 25 years, while in Kenya, it is one of the worst in 70 years (FSNWG, 2020). For Ethiopia, Kenya, and Somalia, grasslands and farmlands were already damaged, with possibly significant implications for such area, in which almost 12 million people have to face extreme food shortages and several dependent on farming for life (FAO, 2020b).

Across the East and African continent, the biggest DL epidemic in 25 years had resulted in major crop production losses, primarily in different agricultural regions of eastern Ethiopia, central Somalia, and northern Kenya (IPC, 2020a). That DL affected roughly 806,400 agricultural families, 197,163 acres of agriculture, and 1,350,000 acres of grassland or browsing (50% pasture damage) within the region, resulting in approximately grain losses of 3,562,856 tonnes (356,286 MT) (FAO et al., 2020). Consequently, as a result, an average of 20 million inhabitants in Ethiopia,

South Sudan, Kenya, Somalia, Uganda, Rwanda, Burundi, Djibouti, and Eritrea are estimated to be vulnerable to food insecurity (WFP, 2020a).

Furthermore, Ethiopia cleaned 700,000 acres in January 2020, Kenya 20,000 acres, Sudan 18,714 acres, Eritrea 15,068 hectares, and Somalia 15,000 hectares. The East and African continent area are now experiencing the worst of all DL outbreaks, with widespread crop destruction or worsening food shortages for nations currently suffering from repeated droughts, war, and rising food prices (IPC, 2020a). Around ten million people across Ethiopia, Kenya, Somalia, and Sudan are already suffering serious food scarcity crises (IPC Phase 3) or worse because of the desert locust invasion. Swarms were threatening an additional 3.24 million of people in Uganda and South Sudan (IPC, 2020a).

That DL invasion is occurring in the middle of a series of climatic disasters, particularly the 2018–2019 drought, which has severely degraded household capacity to cope with in area, left an approximately 18–22 million population suffering a catastrophe as food insecurity worsens (IFRC, 2020a). With DL-affected regions, food production declines are often severe among impacted people. Losses in the impacted areas might exacerbate food shortages, especially mostly in the presence of numerous disasters and high susceptibility (FSNWG, 2020). Scared that the swarms might spread to additional nations in East Africa if "efforts to deal with the voracious pest are not scaled up across the region," the UN agency called for a coordinated response to the disaster (UN news, 2020).

12.4 IMPACTS OF COVID-19 CRISIS ON LIVES AND FOOD SECURITY IN DEVELOPING COUNTRIES

COVID-19 is already spread in poor nations, posing a serious threat to people's health and income. Its most critical and essential part of long-term development is agriculture. Agriculture plays a vital role of the economy. It provides a basic income for a huge part of the population in underdeveloped nations. As a result, disruptions in agricultural production and agribusiness would have far consequences for such regions. Because of the significance of both sectors, this article examines the impact of COVID-19 on food production and agriculture in depth. This study provides trying to cope and mitigating strategies that might be used to maintain one's income.

Deathly health disasters and epidemics have plagued this planet for centuries. The majority of such crises are also the outcome of an exceptional tragedy that has caused individuals to make sacrifices in their current lives. This COVID-19 epidemic is worsening with each day, causing disruptions in anthropogenic activities, an increased mortality toll, or a significant economic impact. COVID-19's causing destruction is not surprising, according to Saqr and Wasson (2020), because epidemics such as the Spanish flu, Polio, HIV, Severe Acute Respiratory Syndrome (SARS), Malaria, the Middle East Respiratory Syndrome (MERS), and Ebola are the current reality of a contemporary world.

According to Adda (2016), total number of people who died from virus infection mostly in the previous century was significantly greater than the number of people who died in large violent clashes. For example, around the turn of the century, the Spanish flu became one of the most devastating epidemics in history, causing the deaths of 50 million people globally. Past epidemics had lower dangers since migration and urbanization have not been as prevalent as they are now, and hence the illness cannot spread as quickly. A new Corona Virus Disease, subsequently dubbed "COVID-19" only after World Health Organization (WHO) classified that as an epidemic, was first identified in Wuhan, China, in December 2019.

Since that day, it has become a pandemic that has spread throughout china. Following then, other instances in adjacent countries such as Thailand, Japan, Singapore, and South Korea were verified. COVID-19 was considered an epidemic and a worldwide alarm 6 weeks after that case was identified by the World Health Organization. The epidemic, unfortunately, expanded rapidly and posed a problem for medical institutions, particularly in southern Europe and North America. Unable to learn through prior tragedies, a lack of urgency in responding to warning people by health organizations and researchers, globalization, and highly linked organisms living are all seen to be factored in the epidemic's spread and effects Saqr and Wasson (2020).

According to estimates, more than 28 million reported cases, and more than 20.5 million recovered, and much more than 900,000 fatalities were documented worldwide within the second week of September 2020, with the United States having the most, then by India, Brazil, and Russia. Even as the world develops increasingly interconnected as a globalized world, many nations are much more prone to be impacted by this health catastrophe. As a result, if an epidemic is not prevented inside one nation, then entire planet would be at risk. The pandemic's impact seems to have a detrimental impact not only on the healthcare systems, but also on severe financial, social, including political issues that would leave long marks when it is not handled in quickly. That condition was grave, and it is compounded in emerging economies, in which it is hard to fulfill the fundamental subsistence demands of such majority of the people. According to a study conducted through the United Nations Sustainable Development Goals (UNSDGs), 55% of the world population, also with the bulk of people living in poor countries, lacks access to basic services. Such susceptibility would exacerbate socioeconomic damages, with implications for human rights, for example, along with denying the poor and disadvantaged of food and basic nourishment. In those civilizations, COVID-19 preventative methods like as closure, stay-at-home orders, general isolation, and transportation stop all extremely difficult. Once the first case was verified, several major Magazine Pre-proof Journal Pre-proof countries declared a brief state of emergency. Imports and exports are getting delayed, if not completely halted in certain circumstances. During the recent economic crisis, countries must rely on emergency funding that may not be sufficient to protect them with their most vulnerable people in society. Due to such lockdown enforced in some nations, the transportation industry also has ceased operations, disrupting the production process for critical products, including food and human aid supplied by various organizations. Even though the extent to which these procedures are implemented varies, they had a significant impact of COVID-19 control. Unfortunately, there is no thorough evaluation of the pandemic's effects in underdeveloped countries. Considering those realities, we focused on COVID-19's implications on sustainability, notably in terms of food security and agriculture in underdeveloped countries. The writers also sought to offer potential mitigating and coping techniques.

12.5 DURING THE COVID-19 PANDEMIC, THE RESPONSE TO FOOD SECURITY WAS CRITICAL

Both national and global food systems have been jolted by the COVID-19 pandemic's effects. This jolt is mostly the result of insufficient preventative measures, or its features and severity would be determined with how administrations and communities react to it and prepare for the disaster. As in lacking of meaningful data and information, it is challenging to make appropriate choices or give help to vulnerable populations, especially when governments lack the resources to assist on food management control. An epidemic can immediately affect the supply of food and consumption circuits, which suggest a fall in excess food and also an increase in food costs. Whenever the epidemic spreads, that will have a negative impact on purchasing ability or the capacity to make and distribute food. Later, on either hand, may differ in terms of severity and it will disproportionately impact the weak (most notably women, the elderly, and children) and the impoverished.

12.6 COVID-19'S IMPACT ON THE AGRICULTURE INDUSTRY

Agriculture is the main source of several developing countries, providing livelihood to over 1 billion people throughout the world. That pandemic's prevention attempts are obstructing the production and supply of agricultural commodities. The agricultural output would be a lengthy process that entails effort in numerous phases, through sowing and nursing to harvest to product export. As a result of its market-based supply chain, both food and farming industries are deemed least stable. Restrictions on travel established in various countries throughout the world that prevent the further spread of coronavirus have impeded those efforts. Agriculture is a high-input sector. For agroecosystems, the reliance upon every element of growth can vary dramatically, exposing farmers or growers to increases in input prices.

And those are the consequence of even a variety of causes, like clogged ports or highways, difficulties in clearing customs, transportation disruptions, the scarcity on financing, or rising prices and investment prices, which can contribute to higher product prices. Those added expenditures made agricultural goods production lesser economical, perhaps resulting in a large loss for farmers. Agriculture or farmers are two words that come to mind while thinking about farming. Insecticides were in short supply inside the nation's first hit, complicating attempts to preserve crops. As a consequence, this year's harvests will be lower. The cost of transporting herbicides to countries such as eastern Africa has increased by nearly thrice. It has slowed locust control efforts thus exacerbating the danger to crop production. Even as the growing season comes, restrictions on trade and import interruptions may impact farmers.

Resilience techniques suggested the mentioned facts in poor nations necessitate prompt attention to mitigate the pandemic's negative consequences. While a major setback, a possible roadmap is urgently needed for every developing nations to always be on pace toward meeting that Goals' established objectives by 2030. As COVID-19 remains and reappears, then following alternative actions could be used to mitigate the pandemic's consequences.

- Food Aid Program: This is critical to strengthen social and food security programs, especially in rural areas, is order to distribute foodstuffs or other requirements for families suffering significant income declines. Barrett (2020) too has mentioned that the necessity to similar efforts, mentioning the 2008 and 2011 food price surges as examples.
- Cash Transfer: Another alternative but successful survival approach for such food aid program, in which weak groups like migrants, landless workers, expectant women, pro-poor people, or others could be helped with such a set amount of money that satisfies their basic livelihood necessities.
- Distribution Network Management: Even during epidemic, the bulk of underdeveloped countries experienced supply disruptions. As a result, effective measures such as increasing storage capacity for increasing reserve funds, reprocessing, or distribution would aid in order to minimize the negative scenario.
- Using Technology: By utilizing the benefits from Information and Communication Technologies (ICTs), overall distribution network management's functionality and efficiency may be greatly improved. Movable buying of perishable food, for example, might be rewarded.
- Globalizing Food System: Promoting a decentralized food supply chain in developing countries that help them become self-sufficient while also conserving food security. It will aid in the achievement of a better food system, which is an SDG 2 goal.

12.7 CONCLUSION

The locust swarm and COVID-19 both happened together in various parts of the world. These events deadly affected the economy of countries not only in Asia and Africa, but also in other parts of the world. The millions of people were affected by both disasters while agriculture was at the top. In terms of Locust, approximately 50% cropland damaged and loss reported in billion dollars while in terms of COVID-19 pandemic, many farmers face traveling ban for their transport of products. As a result of its market—based supply chain, both food and farming industries are deemed least stable. Restrictions on travel established in various countries throughout the world that prevent further spread of coronavirus have impeded those efforts. Agriculture is a high-input sector. For agro-ecosystems, the reliance upon every element of growth can vary dramatically, exposing farmers or growers to increases in input prices. It was concluded that early pre-cautionary measures should adopt at Government and Policy makers to cope with the Locust outbreak and strict measures should be taken the same as like in the COVID-19 outbreak.

KEYWORDS

- **outbreak**
- **pandemic**
- **human health**
- **world war**
- **economy**
- **threats**

REFERENCES

Abaya, S. W. *Floods and Health in Gambella Region, Ethiopia: An Assessment of the Strength and Weakness of the Coping Mechanism*; Thesis–M. Sci; Lund University: Lund, Sweden.

Abera, A.; Belay, H.; Zewude, A.; Gidey, B.; Nega, D.; Dufera, B., Assefa, A. Establishment of COVID-19 Testing Laboratory in Resource-Limited Settings: Challenges and Prospects Reported from Ethiopia. *Global Health Action 2020*, *13* (1), 1841963.

Adda, J. Economic Activity and the Spread of Viral Diseases: Evidence from High Frequency Data. *Quart. J. Econ. 2016*, *131* (2), 891–941.

Barrett, C. B. Actions Now Can Curb Food Systems Fallout from COVID-19. *Nat. Food* 2020, *1* (6), 319–320.

Bashir, M. F.; Ma, B.; Komal, B.; Bashir, M. A.; Tan, D.; Bashir, M. Correlation Between Climate Indicators and COVID-19 Pandemic in New York, USA. *Sci. The Total Environ. 2020*, *728*, 138835.

Beyene, L. M.; Ferede, T.; Diriba, G. The Economy Wide Impact of the COVID-19 in Ethiopia: Policy and Recovery Options, 2020.

Bodamaev, S.; Tuwo, A. D. *Indonesia: COVID-19 Economic and Food Security Implications*; World Food Program: Jakarta, 2020.

.; Catherine, S. The Mid-Year Review of the 2020 Humanitarian Response Plan, United Nations Ethiopia. https://ethiopia.un.org/en/89156–2020-ethiopia-humanitarian-response-plan-mid-year-review-released

Chakraborty, I.; Maity, P. COVID-19 Outbreak: Migration, Effects on Society, Global Environment and Prevention. *Sci. Total Environ. 2020*, *728*, 138882.

CNBC. Weather and Natural Disasters, February 2020. https://www.cnbc.com/2020/02/20/plague-of-locusts-threatens-east-african-economies-as-un-sounds-alarm.html [Google Scholar].

COMENSA. Macroeconomic Impacts of COVID-19 in Sub-Saharan- Africa, Special Report, June 5, 2020. https://www.comesa.int/wp-content/uploads/2020/06/Macroeconomic-Impacts-of-COVID-19-in-Sub-SaharanAfrica.-3.pdf

Cressman, K. Desert Locust. *Biol. Environ Hazards Risks Disasters* 2016, 87–105.

Devereux, S. Social Protection Responses to COVID-19 in Africa. *Global Soc. Policy*, 14680181211021260.

Dowlatchahi, M.; Ahmed, M.; Cressman, K. *Desert Locust Situation in Pakistan*; FAO in Pakistan: Islamabad, 2020. http://www.fao.org/pakistan/resources/in-depth/desertlocust-situation-in-pakistan/en/.

Economic Adviser's Wing, Ministry of Finance. Monthly Economic Update. March. Islamabad, 2020.

FAO, Ethiopian Ministry of Agriculture, FEWSNET & Partners. Impact of Desert Locust Infestation on Household Livelihoods and Food Security in Ethiopia. Joint Assessment Findings, April 2020.

FAO. The Worst Desert Locust Outbreak in Decades Threatens Food Security Across East Africa, GIEWS Special Alert, 29 Jan 2020a.

FAO. FAO Appeals for Urgent Support to Fight Worsening Desert Locust Upsurge in the Horn of Africa, Rome, and 30 Jan 2020b. http://www.fao.Org/news/story/en/item/1259082/icode/

FAO. Desert Locust Upsurge—Progress Report on the Response in the Greater Horn of Africa and Yemen, Jan–April 2020c.

FAO. Impact of COVID-19 on Agriculture, Food Systems and Rural Livelihoods in Eastern Africa: Policy and Programmatic Options, July 2020d. https://reliefweb.int/report/burundi/impact-covid–19

FAO. Locust Hub, 2020e. https://locust-hub-hqfao.hub.arcgis.com/ (accessed 8 Sept 2020).
FAO. Addressing the Impacts of COVID-19 in Food Crises, April–Dec 2020f.
Fawzi, I.; Qurani, Z.; Rahmasary, N. *COVID-19: Implication to Food Security*; Tay Juhana Foundation: Jakarta Utara, Indonesia, 2020.
FEWS NET. 2019 Short Rains in East Africa Among the Wettest on Historical Record. Famine Early Warning Systems Network (FEWS NET), January 29, 2020a. https://fews.net/east-africa/specialreport/january–29–2020
Financial Express India, Secondary Agriculture: The Shift Indian Farming Needs. https://www.financialexpress.com/opinion/secondary-agriculture-the-shift-indian-farmingneeds/1807044/, 2020 (accessed 02 September 2020).
FSNWG. East Africa Desert Locust and Food Security Update: Current Upsurge Threatens Upcoming 2020 Agricultural Season, Special Report, 17 Feb 2020.
GAM. East Africa 2020 Flood Impacts on Agriculture, Updated May 19th, 2020. https://reliefweb.int/sites/reliefweb.int/files/resources/Special_Report_East_Africa_202005.pdf
Herring, E.; Campbell, P.; Elmi, M.; Ismail, L.; Jama, J.; McNeill, S. et al. COVID-19 and Sustainable Development in Somalia/Somaliland. *Global Security: Health, Sci. Policy 2020*, *5* (1), 93–110.
HLPE. Impacts of COVID-19 on Food Security and Nutrition: Developing Effective Policy Responses to Address the Hunger and Malnutrition Pandemic. *Committee on World Food Security High Level Panel of Experts on Food Security and Nutrition*, 2020.
Ho, A.; Dascalu, I. Global Disparity and Solidarity in a Pandemic. *Hastings Center Report 2020*, *50* (3), 65–67.
Huang, C.; Wang, Y.; Li, X.; Ren, L.; Zhao, J.; Hu, Y. et al. Clinical Features of Patients Infected with 2019 Novel Coronavirus in Wuhan, China. *Lancet 2020*, *395* (10223), 497–506.
Hurnik, J.; Kober, C.; Plotikov, S.; Vavra, D. Socio-Economic Impact Analysis of COVID-19, Aug 2020.
IFRC. East Africa: Desert Locust and Food Security and Livelihood Situations. *Inf. Bull* Feb 7, 2020. https://reliefweb.int/report/somalia/east-africa
IFRC, E. A. Red Cross Raises the Alarm Over a "Triple Menace" of Floods, COVID-19 and Locusts, 2020b.
IPC. East & Horn of Africa: IPC Food Security Phase Classification and Desert Locusts, 20 Feb 2020. https://reliefweb.int/report/south-sudan/east-horn
Jamal, S. How Will COVID-19 Affect Pakistan Farmers, Food System? *Gulf News 2020, 1*.
Josephson, A.; Kilic, T.; Michler, J. D. Socio-Economic Impacts of COVID-19 in Four African Countries, 2020.
Khan, N.; Fahad, S.; Naushad, M.; Faisal, S. COVID-2019 Locked Down Impact on Dairy Industry in the World, 2020. SSRN 3616325.
Kimani, J.; Adhiambo, J.; Kasiba, R.; Mwangi, P.; Were, V.; Mathenge, J.; ... & Lorway, R. The effects of COVID-19 on the health and socio-economic security of sex workers in Nairobi, Kenya: Emerging intersections with HIV. *Global Public Health*, *15* (7), 1073–1082.
Kimani, J.; Adhiambo, J.; Kasiba, R.; Mwangi, P.; Were, V.; Mathenge, J. et al. The Effects of COVID-19 on the Health and Socio-Economic Security of Sex Workers in Nairobi, Kenya: Emerging Intersections with HIV. *Global Public Health 2020*, *15* (7), 1073–1082.

Kumar, A.; Padhee, A. K.; Kumar, S. How Indian Agriculture Should Change After COVID-19. *Food Secur. 2020*, *12* (4), 837–840.

Laborde, D.; Martin, W.; Swinnen, J.; Vos, R. COVID-19 Risks to Global Food Security. *Science 2020*, *369* (6503), 500–502.

Latif, A.; Niazi, S. COVID-19 Lockdown Sparks Harvest Crises in Pakistan, India. *Anadolu Agency 2020*, *10*.

Mbow, M.; Lell, B.; Jochems, S. P.; Cisse, B.; Mboup, S.; Dewals, B. G. et al. COVID-19 in Africa: Dampening the Storm? *Science 2020*, *369* (6504), 624–626.

Padhee, A. K.; Pingali, P. Lessons from a Pandemic to Repurpose India's Agricultural Policy. *Nature India*, 2020.

Pakistan Bureau of Statistics. Labour Force Survey of Pakistan 2017–18. Islamabad, 2018.

Pakistan Bureau of Statistics . National Accounts 2019–20 Tables. Islamabad, 2020. http://www.pbs.gov.pk/sites/default/files//tables/nationalaccounts/Table-1.pdf

Pakistan Bureau of Statistics. Population Census, 2017. Islamabad, 2017.

Pescaroli, G.; Alexander, D. Understanding Compound, Interconnected, Interacting, and Cascading Risks: A Holistic Framework. *Risk Analys 2018*, *38* (11), 2245–2257.

.; Raga, S. Economic Vulnerabilities to the Coronavirus: Top Countries at Risk. *Overseas Development Institute (ODI) 2020*, *5*.

Rahaman, M.; Saha, O.; Rakhi, N. N.; Chowdhury, M. K.; Sammonds, P.; Kamal, A. M. Overlapping of Locust Swarms with COVID-19 Pandemic: A Cascading Disaster for Africa. *Pathogens Global Health 2020*, *114* (6), 285–286.

Reardon, T.; Mishra, A.; Nuthalapati, C. S.; Bellemare, M. F.; Zilberman, D. COVID-19's Disruption of India's Transformed Food Supply Chains. *Econ. Polit. Wkly 2020*, *55* (18), 18–22.

Sachs, J.; Schmidt-Traub, G.; Kroll, C.; Lafortune, G.; Fuller, G. Sustainable Development Report 2019. *New York: Bertelsmann Stiftung and Sustainable Development Solutions Network (SDSN) 2019*, 30–31.

Salih, A. A.; Baraibar, M.; Mwangi, K. K.; Artan, G. Climate Change and Locust Outbreak in East Africa. *Nat. Clim. Change 2020*, *10* (7), 584–585.

Saqr, M.; Wasson, B. COVID-19: Lost Opportunities and Lessons for the Future. *Int. J. Health Sci. 2020*, *14* (3), 4.

Schmidhuber, J.; Pound, J. Q. B.*COVID-19: Channels of Transmission to Food and Agriculture*, 2020.

Siddiq, A.; Basher, A. Dysfunctional Horticulture Value Chains and the Need for Modern Marketing Infrastructure: The Case of Viet Nam, 2019.

Sumner, A.; Ortiz-Juarez, E.; Hoy, C. *Precarity and the Pandemic: COVID-19 and Poverty Incidence, Intensity, and Severity in Developing Countries* (No. 2020/77). WIDER Working Paper, 2020.

UN News. East Africa Locusts Threaten Food Security Across Sub-Region, Alerts UN Agriculture Agency, 20 Jan 2020.

UNCTAD. The Covid-19 Shock to Developing Countries: Towards a 'Whatever it Takes' Programme for the Two-Thirds of the World's Population Being Left Behind, 2020.

United Nations Development Programme. COVID-19 Pakistan: Socio-Economic Framework, 2020. https://www.pk.undp.org/content/pakistan/en/home/library/development_policy/covid-19-pakistan--socio-economicframework.html.

USAID. East Africa-Desert Locust Crisis. Fact sheet #7, September 30/2020.

WB. (2020). Tackling the Impacts of COVID-19 Is Imperative to Ethiopia's Journey to Prosperity, 29 Oct 2020.

WFP. WFP EAST AFRICA External Situation Update No. 1 July 2020a.

WFP. WFP EAST AFRICA Update on the Deseret Locust Outbreak, June 12, 2020b. https://reliefweb.int/report/ethiopia/wfp-east-africa

World Health Organization. World Health Organization Coronavirus Disease (COVID-19) Dashboard. World Health Organization, 2020.

Index

E

F

G

H

I

J

L